Iomara Barros de Sousa

Geotechnologies and Multimedia Resources in School Cartography

Iomara Barros de Sousa

Geotechnologies and Multimedia Resources in School Cartography

A teaching tool for Primary School II

ScienciaScripts

Imprint

Cover image: www.ingimage.com

This book is a translation from the original published under ISBN 978-3-330-99696-0.

Publisher:
Sciencia Scripts
is a trademark of
Dodo Books Indian Ocean Ltd. and OmniScriptum S.R.L publishing group

120 High Road, East Finchley, London, N2 9ED, United Kingdom
Str. Armeneasca 28/1, office 1, Chisinau MD-2012, Republic of Moldova, Europe
Managing Directors: Ieva Konstantinova, Victoria Ursu
info@omniscriptum.com

Printed at: see last page
ISBN: 978-620-8-60812-5

Index

Presentation

Speak a new language, so that the world will be a new world.
Mawlana Jalal-ad-Din Muhammad Rum

As a result of the dedication to teaching over nine years in the public education system in municipal and state schools in the state of Rio de Janeiro and the search for new ways to teach cartography through academic research, this book, the result of a master's dissertation defended in 2014 at the State University of Rio de Janeiro, presents a methodology for teaching maps with the use of geotechnologies and multimedia resources applied to environmental education in primary school.

In the midst of the free availability of satellite images and free and open source Geographic Information Systems on the internet, we set ourselves the challenge of writing this book, taking care to show researchers and primary school teachers that geotechnologies are tools applied to mapping processes and can be used to teach maps, as long as they are used through teaching methodologies.

We endeavour to show how the re-signification of mapping technologies, under didactic mediation, turns these tools into valuable instruments for representing and perceiving geographical space, taking the student's spatiality as a starting point.

Bringing spatial science and geographic information closer to primary school geography classes is at the heart of this book, with the aim of promoting greater democratisation of spatial information. The aim is to enable teachers to learn about technological resources that can be used to teach maps in their teaching activities, with a view to improving cartography teaching in the public sector.

In this sense, the theme of this book, "Socio-environmental Perception", seeks to awaken critical thinking and awareness among students and teachers of basic education about the importance of actions in favour of preserving the environment through cartographic representation.

This book is theoretically and methodologically based on successful experiments using geotechnologies carried out by researchers in Brazil and in different countries. I begin by contextualising cartography and geotechnologies applied to the teaching of maps to schoolchildren and summarise the chapters that make up this educational work.

In Chapter 1, **Pedagogical practices and the use of spatial representation in Environmental Education**, I tried to relate Environmental Education and Cartography in the second segment of Primary Education. Didactic mediation for the construction of cartographic knowledge was addressed. I discussed the technologies applied to cartography as teaching tools in geography classes without disregarding cartographic representations on paper.

In Chapter 2, **New Technologies Applied to Cartography in Geography Teaching,** I dealt with the different technologies used in the construction of a digital teaching methodology for school cartography. Multimedia cartography, the Global Navigation Satellite System (GNSS), the Virtual Globe (Google Earth) and online mapping platforms were discussed.

In Chapter 3, **Mapping My River, I** presented the digital teaching methodology called Mapping My River (MMR). This teaching tool is divided into three modules: Module 1: Applying new digital technologies to cartography, Module 2: Exploring the Alcântara River in São Gonçalo/RJ and Module 3: Socio-environmental perception of the Alcântara River in São Gonçalo/RJ.

Chapter 4, **Analysis of the maps and multimedia produced by** the students, analyses the maps and multimedia produced by the students as a result of the application of the MMR in two seventh-grade classes from the São Gonçalo/RJ municipal public school system.

We are delighted to invite geography teachers, school cartography researchers and all those who understand the importance of maps for thinking geographically about their living space and different places to read this book.

Happy reading!

INTRODUCTION

The improvement of the Internet and the advent of geotechnologies such as Remote Sensing, the Global Navigation Satellite System (GNSS)[1] and the Geographic Information System (GIS)[2] have enabled the development of new methods, techniques, materials and instruments applied to the mapping process, which contribute to the production of more accurate maps that are produced in less time.

Technological changes have created new opportunities to represent the earth's surface in digital form, creating new relationships between the reader and the map.

In the educational field, research carried out on geoinformation in the teaching of cartography in Brazil and other countries has pointed to the contributions of GPS, remote sensing and the Geographic Information System (GIS) as important materials for teaching maps in basic education, without disregarding other forms of cartographic representations.

The lack of cartographic materials to work with students' spatiality and at other spatial scales is a problem that affects public schools. On the other hand, we are seeing the increasing availability of geoprocessing applications, programmes and software on the internet, showing different possibilities to be explored in the teaching and learning process of Geography in order to work with Cartography.

The aim of this book is to present a methodology for teaching cartography in a digital environment, aimed at the seventh year of primary school in the São Gonçalo/RJ municipal public school system, based on the use of geotechnologies

1 The corresponding acronym in translation is: "Global Navigation Satellite System (GNSS)." GPS is an instrument for navigation and positioning on the earth's surface that works using the system.

2 The corresponding original acronym in translation is: "Geographic Information System (GIS)".

and multimedia resources, called "Mapping".

Meu Rio (MMR)" available at http://www.mapeandomeusrios.com.br. The theme of this teaching tool was the Socio-Environmental Perception of the Alcântara River, located in the municipality of São Gonçalo/RJ.

Fieldwork was carried out as part of the study of the environment, which allowed the empirical to be realised, so that the students collected information through the application of questionnaires and the collection of geographical data for the preparation of a digital map.

We consider this teaching tool to be up-to-date and, in part, it is already part of the contemporary and everyday life of students who are immersed, for the most part, in digital environments, enabling students to think geographically, through maps, about their role as actors in the environment.

CHAPTER 1

PEDAGOGICAL PRACTICES AND THE USE OF SPATIAL REPRESENTATION IN ENVIRONMENTAL EDUCATION

Santos (1985) points out that the process of social production of space must be understood concomitantly in its form, function, structure and process. In this sense, Cartography as a language, a system of communication codes, as Castellar (2005) points out, enables individuals to represent the geographical space on the most diverse spatial and temporal scales, according to their interests and needs, which improves their perception and understanding of geographical phenomena.

We need to demystify the idea that ready-made maps, on paper or digitally, are absolute and unquestionable truths about geographical space.

The following question arises: How has cartography been taught in schools? How does cartographic language contribute to visualising, interpreting and understanding knowledge about the environment?

Professor emeritus Lívia de Oliveira (1977) sees the map as a graphic language for communicating spatial information, essential for students to graphically represent geographical phenomena. In this way, we understand cartography as a language and methodology and no longer as content (CASTELLAR, 2011).

Cartography teaching should not be reduced to the Euclidean metric, as Lévy (2008, p.156) puts it, the "[...] Euclidean approach that is based on a Cartesian view of extension on the idea of an abstract space, independent of the objects that are arranged in it", as there are different graphic languages that allow students to represent the environment with

their subjectivities, values, conceptions and affective relationships with the place.

Pedagogical practices in the teaching of Cartography applied to Environmental Education should consider the student as the protagonist of the teaching-learning process, under the mediation of the teacher, based on activities that contribute to the formation of aware and critical citizens in relation to the environment from their everyday space.

Geographical Education proposes a critical teaching and learning process that, according to Callai (2013), geographical content should have meaning for students' everyday practice and should be worked on through geographical concepts, such as Place. This concept, when worked on in an experimental approach, allows us to understand the process of social production of geographical space at different times in history, in dialogue with other subjects in the school curriculum.

The aim of this chapter is to discuss the relationship between Environmental Education and Cartography in Primary School II, as well as to present the mental map as an important didactic tool for teachers to identify the socio-environmental perception of students in relation to their lived and concrete space, and to discuss how technologies applied to mapping processes can be used in Geography classes without disregarding cartographic representations on paper.

1.1 Environmental education and cartography in the second segment of primary education

Water resources are important sources for life on Earth. In Ancient Greece, Thales of Miletus, a Western pre-Socratic philosopher, stated that water was *physis*, "the original source", the principle of all things that existed on Earth, conceiving water as an essential source of energy for maintaining and sustaining life on Earth (SOUZA, 1996).

Throughout human history, in the search for new spaces to occupy, man has converted nature into a means of production and has not been properly concerned with the rational use of water resources; thus, irregular constructions have been made along riverbeds, works of canalisation, rectification, widening, without considering their geological, geomorphological and hydrological aspects, as Botelho and Silva (2011) point out.

Although Brazil has legislation on water resource policies, the socio-spatial organisation of Brazilian cities throughout the 20th century was not, for the most part, accompanied by public policies that sought to reconcile urbanisation and the preservation of water quality, and most of the time a rationalist view prevailed rather than a preservationist conception of water resources.

The first steps related to debates on the environment in Brazil, despite legislation aimed at the protection and rational use of natural resources, only occurred with the promulgation of the 1988 Constitution, which, in Chapter VI "On the Environment" of the Constitution (BRASIL, 1988, art. 225, § 1) we find the concern to, "[...] promote environmental education at all levels of education and public awareness for the preservation of the environment".

With the drafting of the National Curriculum Parameters (BRASIL, 1998), environmental issues became part of the Basic Education curriculum in Brazil, as an interdisciplinary theme; this contributed to the study of the environment through different disciplines (Geography, Maths, Science), as stated in the National Curriculum Parameters - Transversal Themes (BRASIL, 1998, p. 193), "[...] students need to build a global vision of environmental issues. 193), "[...] for students to build up a vision of the global nature of environmental issues, it is necessary for each teaching professional, even those specialising in a particular area of knowledge, to be one of the agents of the interdisciplinarity that the subject requires".

According to the National Environmental Education Policy (Brazil, 1999, Law 9795, Art. 10 § 1), "Environmental education should not be implemented as a specific subject in the teaching curriculum", nor should it be seen as a "supplementary chapter", as Porcher, Ferrant and Blot (1975) point out.

Understanding geographical phenomena is closely related to Cartography, an important graphic communication language for graphically representing the complex and dialectical relationships of geographical space by means of a set of signs that have an order, i.e. real information, a visual response to phenomena and must answer the three basic questions proposed by Bertin and discussed by Fonseca and Oliva (2013, p.112):

> 1. In any place, what is there?; 2. Any phenomenon, where is it? (...); 3. For a represented phenomenon, what is its spatial distribution? And more sophisticatedly: In this spatial distribution, what is the order, or what are the proportional relationships?

Research in Brazil in the area of school cartography was initiated by Oliveira (1977, p. 57) where the teaching of maps "[...] must be inserted into the general process of mental development" and the same author adds that "The map is defined, in education, as a visual resource that the teacher must use to teach Geography and that the student must manipulate in order to learn about geographical phenomena".

Environmental education developed in the teaching of cartography must take an interdisciplinary approach involving historical, economic and ecological issues (DEBESSE-ARVISET, 1974, p.17) and therefore contribute to the formation of students "[...] prepared to understand the interdependencies that link life to its environment".

Tristão (2008) emphasises the importance of interdisciplinarity in classroom activities involving the environment through a networked conception of knowledge, i.e. Geography maintains its object of study, which is geographical space, while at the same time cooperating with other subjects in the school curriculum to understand socio-environmental issues.

The praxis of Environmental Education in Geography teaching should consider the student as an active subject in the teaching and learning process, based on the local environment, say Pontuschka, Paganelli and Cacete (2007). In this way, the study of the environment should contribute to the construction of a socio-environmental perception, in which the natural dimension and the social dimension are treated in an integrated way, in a holistic conception, favouring behavioural and attitudinal changes in students.

> It is necessary to take into account not only the multiple dimensions of socio-environmental realities, but also the different dimensions of the person who comes into relationship with these **realities, the globality and complexity of their** "being-in-the-world". **The meaning of "global" here is very different from "planetary";** rather, **it** means holistic, referring to the totality of each being, of each reality, and to the network of relationships that unite beings

> with each other in sets where they acquire meaning. (SAUVÉ, 2005, p. 27)

There is no doubt that environmental issues in Geography lessons are important for building critical and reflective knowledge about the social production of geographical space. To this end, the cartographic language applied in Environmental Education should not be restricted to the Euclidean metric, as Fonseca and Oliva (2013, p.138) state "[...] Euclidean space is only one way of apprehending space" among other possible metrics for building environmental knowledge.

According to Fonseca and Oliva (2003), cartography is a graphic communication language that makes it possible to represent the lived space endowed with subjectivities, meanings and impressions, starting with:

> Place is the basis for the reproduction of life and can be analysed by the *triad inhabitant - identity - place. The* city, for example, is produced and revealed at the level of life and the individual. This plane is that of the place. The relationships that individuals maintain with the spaces they inhabit are expressed every day in the ways they use them, in the most banal conditions, in the secondary, the accidental. It is the space that can be felt, thought about, appropriated and experienced through the body. (CARLOS, 2007, p. 17)

Our harmonious relationship with nature is increasingly limited due to the lack of, as Tuan (1980, p.110) writes, "[...] unconscious gentle involvement with the physical world that occurred in the past [...]" making the environment a symbolically negative "natural" element that is not only a result of the visual and mental, but also includes the other senses. One of the consequences of the disorderly and unbridled development of cities for the environment has been the deterioration of river channels that were once part of people's daily lives, such as using the river for leisure and fishing; as a result, rivers have come to be used as rubbish dumps and perceived as a "ditch".

Using environmental education to work on cartography through the students' everyday experience allows them to identify environmental problems, from graphic representations, as arising from the actions of man and nature, as Martinelli (1994) analyses, in a dynamic and complex dimension.

The use of different forms of representation of environmental phenomena, whether by means of a sketch, plan, thematic chart or mental map, should lead students to go beyond the location of the area under study and thus graphically represent the relationships between the physical and socio-economic environment, contributing to interpretations, explanations and comparisons and, consequently, improving understanding of the interrelationships between man and the environment, encouraging actions and attitudes aimed at protecting and conserving the environment.

> Spatial representation here means more than simply indicating the location of phenomena; it allows us to recover the intelligibility that spatial facts acquire when they are understood from their own contexts (GOMES, 1996, p. 312).

It is important to carry out fieldwork for environmental studies so that students can collect data, seek information about the area under study, make notes, conduct interviews, as Pontuschka, Paganelli and Cacete (2007) point out, in order to improve their perception of the environment based on their lived experience. It's clear to see how cartographic language worked on through the participation and interaction of students contributes to a new way of thinking about man's relationship with nature, leading students to act in favour of preserving natural resources, thus enabling conscious appropriation of natural resources.

1.2 The Mind Map as a teaching tool for studying the environment

Spatial practices, attitudes, beliefs, values and cultural habits attribute subjective meanings to the image of places which, according to Lussault (2003, p. 1), "[...] whether it is material or mental and whether it refers to an objective reality of the physical world or to an abstract ideality" and, therefore, there is great variability in the mental image from one individual to another, although Capel (1973) adds that there are common traits that stand out in a group of people.

From this perspective, the mental map is understood by Kozel (2007, p.115) as "[...] a form of language that reflects the lived space represented in all its nuances, whose signs are social constructions" that represent the individual's perception of the environment, and Richter (2011, p. 125) adds "[...] subjective elements [...] make this representation of one's own hand richer by including contexts that can broaden the understanding of space".

The studies carried out by Gould and White (2002) add that the mental image can be formed either through contact with the place or through the use of media obtained from television, radio or dialogue with another person.

In the current of Humanist Geography, from the perspective of phenomenology, the mental map represents a social, cultural and historical construction as a result of the individual's perception in their multiple senses (touching, hearing, smelling, tasting) of everyday experience with places; it also represents a world view and shows an interpretation of the environment understood in the midst of natural, cultural, historical and social landscapes.

Gould and White (2002) argue that mental maps create possibilities for individuals to represent the real world through the satisfactions, dissatisfactions, needs, values and actions that surround their experiences.

This spatial representation enables teachers to identify the student's mental image formed by their direct and immediate experience and information with the environment. Phenomenology provides support for understanding the socio-environmental relationships that occur in space from the lived world, as Gomes (1996, p. 124) points out, "The lived world is therefore the source and basis of all knowledge and the legitimacy of all consciousness".

Mental maps are representations of spatial perception, without the rigidity of cartographic processes, but they contribute to communication and the reconstitution of real-world information and thus allow the teacher to get to know the students' perception of geographical space which, according to Richter (2011, p. 128), allows them to "[...] include their particular and/or collective interpretations", their perceptions, apprehensions, feelings, becoming part of the cartographically represented geographical space.

Nogueira (2010, p. 129) adds that mental maps make it possible to "[...] know what each person brings with them and what is acquired in their life relationship with the place".

> These maps are not cartographic representations subject to the cartographic rules of projection, scale or precision, but spatial representations originating from the human mind, which need to be read as mappings (processes) and not as static products (SEERMANN, 2003, p. 222).

The article published by Teixeira and Nogueira (1999) on perception and spatial representation contributes to research into mental maps in the teaching of Cartography; this work investigates the mental image formed by students in the 3rd° period of the Geography course at the Federal University of Paraná (UFPR) about the campus where they were studying Geography. The researchers came to the conclusion that the mental map is the result of the mental image constructed from experience and spatial perception.

Staszak (2003) discusses the mental map as a form of cartographic language for understanding individuals' interrelationships with the environment through their cognitive relationship with geographical space.

Thus, the use of mental maps allows students to move from the signified to the signifier, in other words, to represent environmental issues in the space they live in and, in this way, to encode (interpret the map) and decode (translate the map) spatial information and, consequently, to understand phenomena on the scale on which they are perceived and not based on purely mathematical reasoning (CASTRO, 2000). According to Kozel (2010, p.2) "mental maps as symbolic constructions require interpretation/decoding, remembering that they are inserted in collective social, spatial and historical contexts referenced by particularities and singularities. "

According to Gaspar and Marian (1975), the mental map is an image that each of us has in relation to geographical space and, when applied to Environmental Education, it allows the student to represent the environment known and practised in their daily lives and thus express spatial attitudes and knowledge.

Unlike the "naturalism" of ready-made maps that are often subjected to the ideological function of hegemonic powers, the mental map makes it possible to exercise citizenship in order to intervene with a critical eye in the search for solutions and alternatives that allow for dignified coexistence and the well-being of everyone in society, as stated by Reigota (2009).

1.3 Teaching practices and the use of technologies applied to cartography in environmental education

The different and multiple communication and information technologies that permeate students' daily lives, such as computers, smartphones, cameras and the internet, are easily used by school-age teenagers to play games, exchange and receive messages from friends, creating, as Kenski (2004, p.100) describes, "other logics for understanding the world, appropriating information, interpersonal relationships and socialising and participation".

The ease and enthusiasm of students in handling digital technologies allows teachers to work with cartography in a digital environment using satellite images, GPS, GIS and multimedia resources that are present in students' daily lives, but are still far from the classroom.

In this sense, Di Maio's doctoral thesis (2004) on the use of digital geotechnologies in secondary education highlights the potential of geotechnologies in the construction of geographical knowledge, by enabling teachers to carry out cartographic activities with up-to-date spatial information , thus providing students with greater interest, enthusiasm and interaction.

Santos (2002) describes that digital satellite images applied to Environmental Education contribute to a holistic view of the study of the environment and also have a vast potential for use and application in the study of environmental dynamics, given that:

> "**[...] it makes it possible to** identify and relate natural and socio-economic elements present in the landscape such as mountains, plains, rivers, hydrographic basins, forests, arable and industrial areas, cities, etc., as well as to monitor the results of the dynamics of their use, thus serving as an important aid to understanding the relationships between men and their consequences in the use and **occupation of spaces and the implications for nature".** (SANTOS, 2002, p. 6)

Nascimento (2003) in his doctoral thesis "Instruments for the practice of Environmental Education with a focus on water resources" noted the importance of using multimedia resources as materials to support the construction of geographical knowledge based on the experiences that students acquire in formal Environmental Education. To this end, the proposal to produce videos allowed students to obtain information and geographical data to build knowledge about the water issue, particularly the Pântano do Sul catchment area in the municipality of Florianópolis/SC. The population growth and lack of basic sanitation observed in the production of the videos showed that multimedia resources contribute to developing actions and attitudes in defence of drinking water through the proposal made by the students to implement basic sanitation in the municipality and, consequently, improve the apprehension of spatial dynamics.

Sachs (2002, p.31-32) stresses that the study of the environment should include "[...] a holistic and interdisciplinary approach [...] in favour of wise ways of using and exploiting nature's resources, respecting its diversity".

Geotechnologies and multimedia resources used in Environmental Education are shown to be materials that make a great contribution to the construction of knowledge related to the environment, as discussed in the following article

Perrenoud (2000, p.139) "The new technologies can reinforce the contribution of contemporary pedagogical and didactic work, because they allow rich, complex and diversified learning situations to be created," by considering the student as an active subject and the teacher as a mediator of the teaching and learning process.

In this sense, Perrenoud (2000, p.139) states that "[...] more than teaching, it's about making people learn", in other words, pedagogical practices must go beyond teaching content.

The use of geotechnologies and multimedia applied to cartography in the study of the environment should support formal environmental education through reflection and rational awareness, thus enabling students to recognise their responsibilities in relation to the environment to which they belong. Sachs (2002, p.18) shows that "[...] only through teaching can this universal awareness of everyone's responsibility for protecting the biosphere be achieved".

The teaching and learning process in a society increasingly permeated by computerised technology requires the training of teachers who are not just content experts, but above all educators who develop methodologies that encourage student involvement and interaction in the construction of knowledge.

Firstly, it is necessary to consider the immediate environment and, therefore, reflect and criticise environmental dynamics by using data and information collected by the students, articulating different agents who produce urban space and, therefore, correlating the scale of the neighbourhood, municipality and even the global scale, in such a way as to provide the student, as Cavalcanti (2006, p. 17) states, with "[...] full training (cognitive, affective and social) for the exercise of their citizenship".

Therefore, the real contribution of geotechnologies and multimedia resources applied to cartography in environmental education is to train critical citizens who are aware of the rational use of the environment.

In the next chapter, we discuss the mapping technologies applied to cartography teaching, specifically GPS, remote sensing and the online mapping platform used to develop the MMR.

CHAPTER 2

MAPPING TECHNOLOGIES APPLIED TO TEACHING CARTOGRAPHY

The technologies applied to Cartography, such as multimedia (text, photo, video), satellite images, GNSS (Global Navigation Satellite System) and the Geographic Information System (GIS), when used in Geography classes, under the mediation of the teacher using teaching methodologies appropriate to the students' cognitive structure, allow the student to explore and interact with geographic information; however, the use of multimedia resources and geotechnologies is still little explored in Geography classes in both primary and secondary schools.

Wiegand (2005) states that pedagogical practices will need to be involved in new ways of looking at the use of maps, so that the cartographic language used in geography teaching uses analogue maps to complement digital maps on the Internet, allowing students to better understand the complexity of spatial relations between society and nature, based on their lived space, produced space and the transformations of the place where they live.

For this reason, this chapter discusses multimedia cartography, geotechnologies (remote sensing, GNSS and GIS) and online mapping platforms, based on theoretical and methodological references, with a view to their contributions, limits and possibilities for using these tools as materials to support cartography in geography teaching.

2.1 Multimedia cartography

Digital cartography, defined as a form of communication, presentation and spatial analysis of a wide variety of topics and interests of individuals in a dynamic, multimedia, multisensory and multidisciplinary way, enables users to include geographical information according to their interests and needs (TAYLOR, 2005).

This new paradigm in cartography, digital cartography, offers conditions for users to improve their understanding of spatial dynamics through interaction with geographical information, allowing them to "travel" inside maps and thus create new forms of relationships with cartography and, consequently, different uses of the map, as Peterson (2007) states.

Among the elements of Digital Cartography, we find Multimedia Cartography which represents a technological revolution for the mapping process, as stated by Menezes (2003), since it is based on the combination of maps with other media through the computer, such as: text, photo, drawing, video and sound, allowing a better understanding of reality, as pointed out by Cartwright (1999); the basic principles of Cartography have not been substantially modified, on the contrary, new technologies have been applied to the mapping process based on the basic elements of Cartography.

Researcher Suely Aparecida Gomes Moreira (2010), in her doctoral thesis "MULTIMEDIA CARTOGRAPHY: interactivity in cartographic projects", points out that the term multimedia can be understood as multi (various) and media (languages), i.e. it is any combination of text, drawing, sound, photo and video used to convey a message according to the context in which it is used.

For the French philosopher Lévy (1999), the media is the vehicle through which the message is received and has a multisensory character, as it involves perception through different sense organs (sight, touch, hearing, kinaesthesia and smell), as Santaella (2003, p. 25) states, "Now, media are means, and means, as the name implies, are simply means, that is, material supports, physical channels, in which languages are embodied and through which they pass".

According to Peterson (1999), the use of multimedia resources in cartography can be identified by five basic principles: a) the inadequacy of representing spatial dynamics on paper, especially as it does not support changes in spatial resolution;

b) paper maps have higher production and distribution costs, while maps produced on the Internet make it possible to produce and make available maps simultaneously and distribute them in a fraction of a second, as well as allowing "dialogue" with the maps, i.e. the user can change the scale of the map to better visualise geographical phenomena;

c) the use of multimedia in spatial representation enables users to form their perspective of the world and, consequently, improve their basic knowledge of cartography and, at the same time, produce a mental representation of their living space, especially space beyond their direct experiences at other spatial scales;

d) the intrinsic value of multimedia: it improves the acquisition of information and the construction of spatial knowledge in a participatory process;

e) the moral obligation of cartographic communication to make people understand geographical space and provide access to information for decision-making.

Because of the contributions of multimedia to cartography and its relevance as a tool for geography classes in basic education (Figure 1):

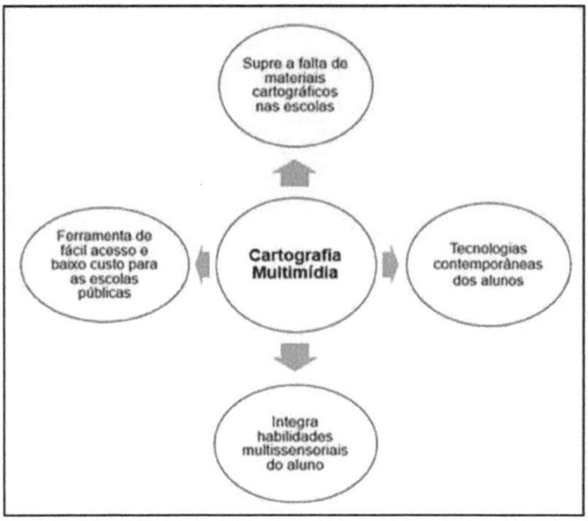

Figure 1 - Multimedia cartography

Source: Prepared by the author Sousa (2017).

Most of the time, cartographic materials in public schools are out of date or do not exist, and when schools do have maps, they are on a small scale, representing a static, unchanging world, and do not adequately fulfil the essence of cartography, which, according to Peterson (1999, p.31), is "the representation and communication of spatial dynamics". Cartographic language using multimedia resources in geography lessons can encourage students to draw up maps according to their needs and interests, including their own experiences.

Multimedia resources to support the teaching of cartography prove to be valuable materials for the teaching and learning process of geography, as they allow the teacher to carry out activities on the student's living space, such as the neighbourhood, district or municipality, expanding to other spatial scales and, above all, offer possibilities for student involvement in the mapping process.

Working with multimedia resources has become an easily accessible and low-cost tool for public schools, since the multimedia resources are obtained by the students themselves from their smartphones or tablets.

obtained free of charge on the internet from their personal computers or from the school's computer lab . The use of multimedia resources in cartography teaching are tools that are easy for students to access and use via their smartphones. Freundschuh and Helleviks (1999) report that studies carried out by Koussoulakou and Krakk (1992), Milheim (1993) and Edgeman (1994) show that students prefer to work with multimedia computer tools because they are more enthusiastic about learning with technology.

Google Earth is an interactive multimedia application available free of charge on the Internet, among other free applications, programmes and software; it allows Geography teachers to work with the production of maps using satellite images as a teaching tool that allows students to extract and insert the most up-to-date information on geographical space. In this way, the teaching and learning process becomes more interactive and stimulating, as the student inserts geographical information using media (photo, video and text), chooses "information planes" such as places, roads, boundaries and borders, uses resources to change the visualisation of the area under study and modifies the scale, according to the interest of the area under study, and also shares their map on the Internet.

Maps integrated with multimedia and the internet allow geography teachers to work with contemporary electronic media such as computers, the internet and smartphones, which makes the teaching and learning process more attractive, motivating and enthusiastic, as students have the opportunity to develop their spatial representation in the classroom using tools from their everyday lives.

The use of multimedia tools in map teaching makes it possible to develop multisensory learning, i.e. the combined use of senses such as sight, hearing, touch and smell, facilitating the construction of spatial representation by integrating maps with media obtained by the students. Multimedia resources help work with multiple intelligences (spatial, linguistic, musical, bodily-kinesthetic, logical-mathematical, interpersonal, intrapersonal) which, according to the Theory of Multiple Intelligences proposed by Gardner (1995), allow problems to be solved or products to be produced, in the case of this research, maps. Freundschuh and Helleviks (1999) showed that the use of multimedia in Cartography and Geographical Education develops a critical, creative spirit in students through learning developed from geographical information they have collected.

In this sense, teaching geography using multimedia resources allows students to create a graphic representation as a result of their own experiences. The main contribution of multimedia in cartography is to improve the apprehension of geographical phenomena due to the involvement of multiple intelligences, notably spatial intelligence and kinaesthetic intelligence. Once multimedia tools are associated with the teaching of maps, they can facilitate the graphic representation of the real world and, consequently, the understanding of the geographical fact or phenomenon represented in a more interactive and participatory way.

According to Wolfgram (1994, apud Russo and Garcia, 2001), people remember only 15 per cent of what they hear, 25 per cent of what they see and 60 per cent of what they interact with. This finding justifies the importance of using interactive computer-based materials in the teaching of cartography in order to improve cartographic communication and allow students to select and elaborate their spatial representation and, consequently, make the use of maps more attractive for those who teach and/or learn.

According to Taylor and Lauriault (2007), computer reality is no substitute for the real world, but Multimedia Cartography can help us make geographical communication more interesting, engaging and affective.

2.1.1 Interactivity and Multimedia Cartography

The concept of interactivity existed even before the advent of information technology and refers to communication, not information technology. For sociologist Silva (2001), interactivity is not limited to digital technologies; this term can be used to designate communication between people, between humans and machines and between users and services.

Silva (2001) adds that interactivity depends on two basic positions: dialogue and user intervention in the content of the message.

Although interactivity is increasingly present in cartographic language in digital media, the term interactivity is not new to cartography. Peterson (1998) states that the first map built in the sand, in prehistoric times, resulted from the interaction between two individuals: the interactor and the "cartographer" who included a reference point according to the interlocutor's needs, in other words, the first maps were interactive.

Over time, maps have lost their interactive character and have been characterised by a static and unchanging spatial representation without the user having any involvement in the mapping process, although access has increased, as Ramos (2005) points out.

The application of computer technology to cartographic language has made it possible to recapture the interactive potential of maps, as stated by Ramos (2005), by allowing the user to "converse" with the map, creating new relationships with the spatial representation. Kraak (2001) shows that static maps can be interactive, as long as they function as an interface with other information, other maps or images.

> The interactive map is characterised by an interface made up of graphic icons, a location device and an instant map function. The interactive map includes tools for zooming in or increasing the scale of the map in different areas to insert videos of places with photos and sounds (PETERSON, 1995, p. 45).[3]

This research was based on the concept of interactivity in pedagogical activities in the teaching of cartography and considered interactive multimedia with a non-linear structure (the themes/contents are made available to the student to navigate "freely", but in accordance with the activities proposed and guided by the teacher in the classroom), in line with Vaughan (1994), in which students can "navigate" freely through the maps, according to the scale that best visualises the map.

The use of interactive maps in cartography teaching makes it possible to develop activities through student participation in the production of the map by inserting geographic data and information obtained and collected by the students themselves.

Interactivity in cartographic materials for schoolchildren makes it possible to change the level of detail of the information on the map, i.e. to reduce or enlarge the scale of the map, according to the need to visualise the phenomenon while respecting the student's own pace and pattern of decoding information. This differs from ready-made maps which, more often than not, do not take into account the student's cognitive level in order to decode geographical phenomena.

2.2 Geotechnologies applied to geography teaching

Geotechnologies comprise a set of technologies for collecting, processing, analysing and manipulating geographically-referenced information that makes it

[3] The foreign language text is: "The interactive map is characterised by an intuitive user interface consisting of graphical icons, a pointing device and the near instantaneous display of 26 maps. The interactive map includes "tools" to further zoom in on the map or "open-up" different areas and may include "videoclips" of places with pictures and sound.

possible to "(re)know" the *Earth* at different spatial and temporal scales, as Santos (2002) points out.

In recent decades, the development of modern means of communication/information applied to cartography, such as remote sensing, GNSS and GIS, have made it possible to obtain more accurate, up-to-date and faster spatial data and information on the earth's surface.

In the mapping process, these geotechnologies have proved to be important support materials for cartographic activities in Geography classes, as they allow students to locate, correlate and analyse geographical factors acting on the dynamics of the earth's surface with data/information on different spatial and temporal scales from their living space, such as the neighbourhood.

Researcher José Jésus Reyes Nunes (NUNEZ, 2006) has developed digital material related to the teaching of cartography for primary and higher education students in Hungary. This material, called *'TÉRKÉPTORTÉNET'*, brings together three themes: the History of Maps, which covers a little about the history of Hungary, the Geographical Information System and Multimedia Cartography with interactive texts and cartographic games.

In this context, the doctoral thesis by Prof Angelica Carvalho Di Maio of the Fluminense Federal University (2004) shows the importance of using geotechnologies to teach cartography in geography classes:

> [...] an additional, well-founded instructional tool for learning geography, which involves the study of dynamic processes using digital cartography, remote sensing, the geographic information system and the use of GPS, in other words, the available geotechnologies. (DI MAIO, 2004, p. 135).

Therefore, the use of geotechnologies makes it possible to build and/or solidify knowledge relating to the basic elements of cartography (scale, geographical coordinates, legend, spatial orientation), using tools that permeate students' daily lives and contribute to the formation of participatory, critical citizens who are aware of spatial dynamics. Despite this, these geotechnologies are little explored in the preparation of maps in the classroom.

2.2.1 Global Navigation Satellite System (GNSS)

The term Global Navigation Satellite System (GNSS) was coined in 1991. It has global positioning capabilities and includes GPS, GLONASS and GALILEO, as well as China's CAMPASSS.

GPS was developed in the 1960s by the Navstar project and provides the positioning of a mobile object through the coordinate system, . It measures the distance from a point anywhere in the world with great precision, thus adding greater dynamism to the survey and representation of geographical data (BERNARDI E LANDIM, 2002).

Among the GNSS, GPS (Global Positioning System) is the satellite localisation system most used in the mapping process. According to Bernardi and Landim (2002), GPS consists of three main segments:

a) space - is related to the satellites orbiting the Earth, distributed in 6 different orbits at an altitude of approximately 20,200 kilometres with an inclination of 55 in relation to the equator and a revolution period of 12 hours. This configuration ensures that at least four satellites are captured by GPS, allowing it to be used anywhere on Earth;

b) control - is related to the monitoring and control of the satellite system and updates the navigation data of each satellite, which is carried out at five global monitoring stations;

c) user - is related to the user community, which can be divided into civilian (navigation by cars, buses, use in extreme sports, positioning of objects) and military (training, conflict movements).

Increasingly, GPS is part of the daily lives of teenagers and young people of school age; however, most of them don't know how to interpret and read information about routes, journeys and data about a given area.

The use of this geotechnology in geography lessons allows students to monitor an environmental area by obtaining geographical coordinates (latitude and longitude), spatial orientation, altitude and distance and displacement measurements in the area under study.

The use of GPS allows teachers to develop more interactive pedagogical practices and, consequently, to get students more involved and engaged in the teaching of maps, according to studies carried out by Di Maio (2004), when compared to the use of paper maps, insofar as it creates possibilities for students to obtain, for example, geographical coordinates of a location by simply positioning the GPS on the terrain.

Gómez's (2013) experience with GPS in a seventh grade class shows that this geotechnology can be used to build knowledge through practical experiences in public schools. The students discovered the location of the same symbols on a football pitch using geographical coordinates (latitude and longitude). They also explored spatial orientation, travelling speed and distance and altitude measurements.

2.2.2 Virtual Globes (Google Earth)

Remote Sensing technology is used to support the teaching of Cartography in Geography classes, such as satellite images which, according to Florenzano (2007), are "faithful portraits" of the earth's surface.

Google Earth is a programme available free of charge on the Internet, made up of satellite images and maps, which allows you to move the image around, use tools to zoom in and zoom out, trace paths and calculate distances between places, "walk" through known and unknown locations, record videos, identify changes in landscapes over the last ten years and thus allow anyone to create their own map.

In this sense, the use of satellite images through the Google Earth programme is a valuable tool for cartographic activities in Geography classes. Teachers can work with the most up-to-date information about the geographical space, such as the student's living space, be it a city, a neighbourhood or even a block, enabling students to build knowledge about spatial dynamics in a more interactive and participatory way.

According to Crampton (2010), the use of Google Earth has emerged as a possibility for making maps outside the confines of academia. This new way of building maps, with interactivity in the third dimension, makes this programme a support material for teaching maps in primary schools.

The insertion of this spatial technology offers students the opportunity to go beyond the location of a geographical phenomenon, to carry out analyses and correlations on spatial transformations of a given place, as well as allowing them to change the spatial scale and time scale, move the map, trace a route, record a walk, draw paths and polygons, overlay layers that can contain various spatial data, such as places, visualisation of houses, buildings, terrain elevations in the third dimension.

In addition, students can add information about a given area through texts, photos, videos and, above all, produce their own map and share it on the Internet via Google Earth Community, as Di Maio and Stezer (2011) point out. This favours the construction of broader and more diversified knowledge about spatial dynamics, which often arouses and stimulates greater interest in Geography lessons.

The interactive and trial-and-error nature of Google Earth contributes to the development of a teaching and learning process in which the student becomes a participant in the mapping process by editing their own map:

> Geography teachers and students can visualise many places on the earth's surface from any angle and any distance. These tools simulate flying over the landscape and stopping at the specific point of interest. In any case, you can zoom in, move around or visualise the Earth in the third dimension. (KERSKI, 2006, p.1)[4]

There are major challenges to overcome when it comes to incorporating spatial technology via the Internet into cartography teaching in Brazil's public

[4] The foreign language text is: "[...] the geography educator and student can view most locations on the Earth's surface from any angle and from any distance. The tool simulates flying above the landscape and stopping at key points of interest that you specify. At any point, you can zoom, pan, or tilt the Earth view for a 3D perspective." (KERSKI, 2009, p.1).

primary school system. These include: the infrastructure of computer labs, which is characterised by a disproportion between the number of computers and the number of students in each class, with an average of 35 to 40 students, slow browsing speeds and teachers' lack of knowledge or even interest in the benefits of using Google Earth in the process of teaching and learning geography.

It is necessary to reflect on the use of Google Earth in the teaching of maps in order to improve students' cartographic knowledge or solidify possible gaps that have not been learnt through the use of paper maps.

The inclusion of this programme in Geography classes requires teaching methodologies that contribute to the development of students' cognition and spatial perception, because in some places satellite images are marked by low resolution characterised by the excess or omission of geographical information, as Gonçalves et al. (2007) point out, due to social and political contexts, as Crampton (2010) adds. Novaes (2014) shows the choices that Google Earth makes to represent

often imposes limits to control territories, such as communities; in other places, satellite images show good resolution due to the mechanisms of strategies and control of urban space by state authorities.

Martins, Seabra and Carvalho (2013) pointed out the use of Google Earth for teaching Geography in the construction of a map of the students' neighbourhood of residence for the 6th grade. The results showed the importance of prior knowledge for working with spatial technology in cartographic activities with the students' living space.

2.2.3 *Online* mapping platform - the use of ArcGIS *Online*

In parallel with the availability of GIS installed on personal computers, GIS was developed on the Internet from the 1990s onwards, becoming a powerful tool for mapping and environmental analyses (BACKER, 2015). In the 2000s, platforms were created for mapping on the Internet, such as ArcGIS online[5] ; this, in turn, is based on the cloud geocollaboration system[6] which allows users to include map bases developed in a GIS in WinZip format. This helps to improve interpretation and analysis and makes it less abstract for students to understand spatial dynamics.

ArcGIS *Online* is a Geographic Information System based on the[7] geocollaboration cloud system that allows users to include map bases in WinZip format and use them together with map bases made available in this GIS. This helps to improve the interpretation, analysis and apprehension of spatial dynamics as the user interacts with geographic information.

Santos (2010) found that the use of ArcGIS version 9.3 helps to teach Geography in basic education, promoting improvements in the learning of related content, although this GIS can also be explored by other curricular subjects. The application of this geotechnology made the class more interesting, there was greater involvement and participation by the students in carrying out activities using geotechnologies; this was facilitated because the students are contemporary with digital technologies.

In his doctoral thesis, Hong (2012) discusses the importance of teacher training in geotechnologies for implementation in the classroom, bearing in mind that the use of ArcGIS *online* offers students the opportunity to map, conduct research and analyse spatial information at different spatial scales from their living space from any computer or mobile device such as *smartphones*, thus contributing to meaningful learning of geographical space.

Kerski (2013) adds that this platform is a powerful tool for teaching cartography, since it is possible to integrate a cartographic base with the platform's own mapping base, which includes orbital images, layers and

---

[5] ArcGIS online is a product commercialised by the private company ESRI and available free of charge on the World Wide Web at: < https://www.arcgis.com/home/index.html>.

[6] The cloud system is the operating system of programmes or software stored on servers interconnected via the Internet, taking up no space on the computer's hard drive.

[7] The cloud system is the operating system of programmes or software stored on servers interconnected via the Internet, taking up no space on the computer's hard drive.

topographic maps. In this way, students can create, modify and update their maps, making it easier to understand the dialectical and contradictory production between society and nature.

In this sense, ArcGIS *Online* presents itself as potential material for teaching cartography, since it doesn't cost schools any money, it doesn't take up space on the hard disc, since it can be accessed via the Internet; users just need to sign up for a free account to create their map. In addition, ArcGIS *Online* allows teachers to develop cartography teaching by integrating various contents through an interdisciplinary approach.

The experience of Gomes (2006), in his doctoral thesis, indicates that the use of the Arc Voyager GIS in primary education is a tool with didactic potential for working on topics related to Geography, in the case of this research, "The Earth: studies and representations and Population and Population: Population - behaviour of demographic indicators". The results of the research showed that the use of this geotechnology in Geography teaching helps to improve student performance and, consequently, school performance in the subject of Geography.

It is well known that the digital technologies used by school-age teenagers arouse great enthusiasm and involvement in their daily lives. According to the experience of Di Maio (2004), lessons become more interesting and engaging with teaching methodologies in which the student is a participant in the map learning process, which consequently improves understanding of abstract issues and makes the use of maps meaningful for everyday practices.

Pazini (2004), in partnership with the Geoprocessing Technology Centre, one of the units of the Paulista Foundation (CETEC), developed a methodological proposal for the 3rd and 4th cycles of primary school using geoprocessing in geography teaching. The methodology pointed to positive aspects regarding geotechnology in the study of spatial dynamics, namely: the student's ability to correlate isolated spatial phenomena to a broader context, to draw up maps with up-to-date information, creativity, enthusiasm and greater involvement in cartographic activities.

The inclusion of GIS in basic education makes it possible to work on issues related to the student's living space, which most of the time are not covered in textbooks or atlases because they contain programme content at national level; however, the use of geotechnologies in schools is still very limited due to structural barriers (low number of computers per class, low information processing capacity due to the speed of the Internet connection) and social barriers (teachers' difficulties in attending training courses due to financial conditions and lack of time).

The research carried out by Vilhena, Junior and Beserra (2012) showed that the use of geoprocessing technology is valuable material for teaching Geography. High school students therefore produced an image chart of the main flooding points in the urban perimeter of the city of Boa Vista/RR using the CCD/CBERS image through SPRING 5.1.5. They found that the students had identified and understood the main socio-environmental factors resulting from flooding in the city of Boa Vista, as well as their potentialising agents.

Therefore, this geotechnology in cartography teaching allows students to change the scale of the map by modifying the level of detail to better visualise spatial information, correlate spatial phenomena from different information planes, calculate areas in an interactive teaching and learning process and thus contribute to developing geographical knowledge and improving the quality of cartographic education.

CHAPTER 3

MAPPING MY RIVER

Mapeando Meu Rio (MMR) is a teaching tool for cartography aimed at the seventh year of primary school, via the Internet, using geotechnologies and multimedia resources to study the environment in Geography classes, which is available free of charge at the following email address: http: <//www.mapeandomeusrios.com.br>, shown in Figure 2.

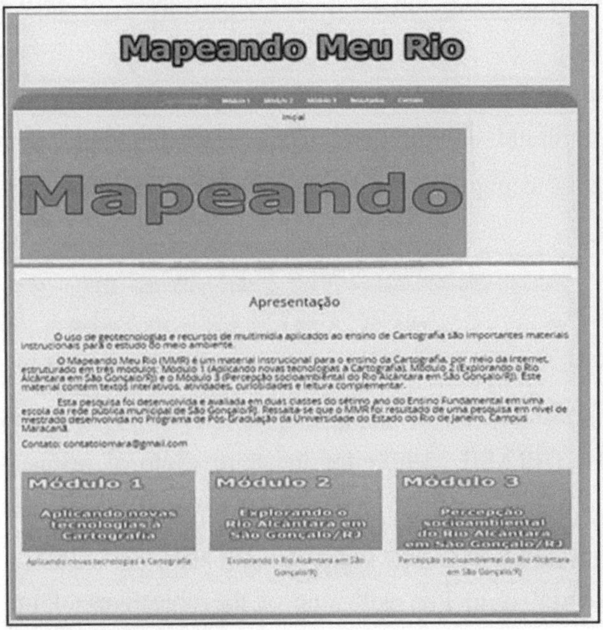

Figura 2 MMR homepage

Source: http://www.mapeandomeusrios.com.br/

The MMR was implemented in an *online* content management system (CMS). Wordpress is an open-source system whose programming language is based on PHP code with a MySQL database.

easy to use, flexible, which allows the teacher to develop teaching methodologies in a digital environment, with the possibility of an exclusive environment for managing all the content and visual structure of the material developed.

The main advantage of using Wordpress in education, especially in the public school system, is that it is completely free. Another positive aspect of this platform is that the administrator, in this case the teacher, can give the students access to the site and allow them to create a *login* and password to insert their activities, under the guidance of the teacher. In this way, students can publish on the World Wide Web their research carried out in the field, as well as their cartographic representations with multimedia resources produced by them which, in the first instance, need to be assessed and accepted by the teacher for *online* access.

The purpose of Mapping My River is to provide a methodology for teaching cartography in a digital environment, using geotechnologies and multimedia resources, helping to improve environmental perception, based on the spatiality of the students.

2.1 Choice of content

The content was chosen based on the Geography PCNs (BRASIL, 1998) and the Environment PCNs (BRASIL, 1998) for the third cycle of primary school, in line with the curriculum proposal of the São Gonçalo/RJ municipal education network for the seventh year of schooling (SÃO GONÇALO, 2008).

In this way, this material is in line with one of the objectives of the Municipal Environmental Policy of São Gonçalo/RJ.

> **"promote the permanent and systematic awareness of the** population and the adaptation of teaching within the principle of awareness and mobilisation, in order to incorporate the principles and objectives of Environmental Education into the school and community (SÃO GONÇALO, LEI 016/2011. Art. 3)".

The theme addressed in the MMR was "Socio-environmental perception of the River Alcântara in São Gonçalo/RJ through geotechnologies and multimedia resources applied to the teaching of Cartography in Geography classes". This material was structured into three modules: Module 1 (Applying new technologies to cartography), Module 2 (Exploring the Alcântara River in São Gonçalo/RJ) and Module 3 (Socio-environmental perception of the Alcântara River in São Gonçalo/RJ), as shown in Figure 3:

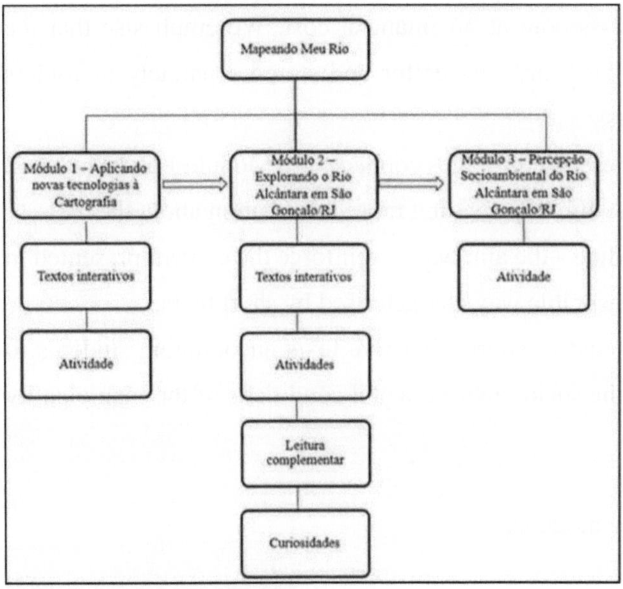

Figura 3 - MMR module structure

Source: Prepared by the author Sousa (2017).

a) activity - in each module there is at least one activity; in Module 1, Google Earth and GPS were used; in Module 2, the activities included a stage in the classroom (drawing up a mental map) and, secondly, fieldwork (interviewing residents and/or traders and gathering environmental and social information on the Alcântara River from one of its channels); in Module 3, the activities were carried out in the classroom (drawing up a mental map) and, secondly, in the field (interviewing residents and/or traders and gathering environmental and social information on the Alcântara River from one of its channels).

we proposed using ArcGIS Online for the students to build their maps. It's important to note that Google Earth allows students to save their steps in the application and continue on to the next lesson; ArcGIS Online allows students to store the completed steps on the Internet without taking up space on the computer's hard drive; thus, these technological resources applied to Cartography are a great contribution to public schools by allowing the insertion of geotechnologies in the classroom at no financial cost. We emphasise that the answers were edited in the Word text editor and stored separately in folders corresponding to each class;

b) interactive texts - these are the pages connected to Module 1 and Module 2 by means of links with the aim of providing more information about the texts;

c) complementary reading - the aim was to reinforce the content presented in a less tiring and more pleasurable way characterised by short texts;

d) curiosities - correspond to short, objective texts encouraging students to learn a little more about the socio-environmental conditions of the municipality of São Gonçalo/RJ.

3.1.1 Division of teaching modules

Mapping My River was structured in three modules, taking into account the cognitive development, basic knowledge of cartography and geographical knowledge of students aged between 11 and 13.

3.1.1.1 Module 1: Applying new digital technologies to cartography

This module dealt with "Axis 4: Cartography as a tool for approaching places *and* the world" as curriculum content for the third level of primary school proposed by the PCNs for Geography (Figure 4).

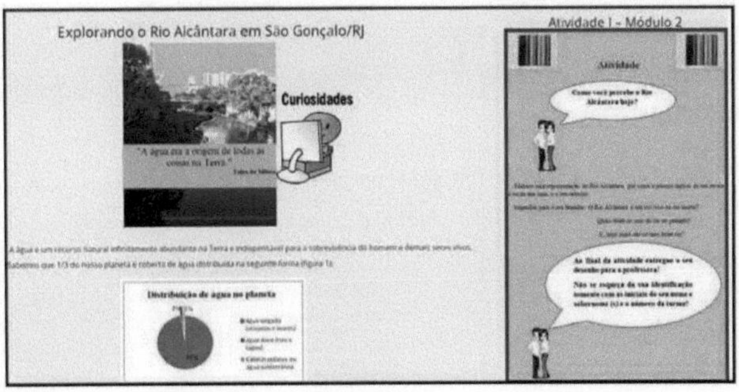

Figura 4 - MMR module 1 screenshots

Source: Prepared by the author Sousa (4).

The activities were developed using GPS and Google Earth (an application available for free on the Internet) in order to solidify the basic knowledge of cartography learnt in the 6th grade° or in previous school years. We used the Paint programme to insert the spatial orientation and draw up the map legend. Table 1 shows the content and objectives of this module:

Table 1 - Content and Objectives of Module 1

Contents	Cartography basics
Objectives	identify the basic elements of cartography (legend, scale and geographical coordinates)
	explore the cartographic categories of localisation, analysis and correlation using multi-temporal satellite images;
	point out the contributions of the new digital technologies used in cartography for teaching using Google Earth and GPS.

Source: Prepared by the author Sousa (2017).

4.1.1.2 Module 2: Exploring the Alcântara River in São Gonçalo/RJ

In this module we begin to develop the theme addressed in Mapping My River: "Socio-environmental Perception of the Alcântara River in the Municipality of São Gonçalo/RJ" which, in turn, is in line with the

curricular content "Axis 2: the study of nature and its importance for man" proposed by the PCN for Geography (BRASIL, 1998) for the third cycle of primary education (Figure 5).

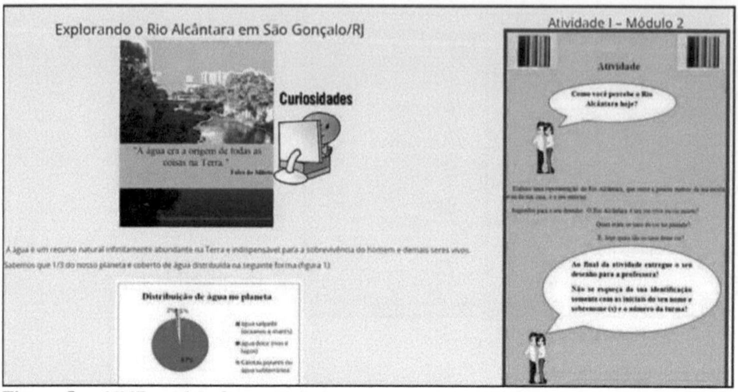

Figura 5 - MMR module 2 screenshots
Source: Prepared by the author Sousa (2014).

The activities in this module comprised the following stages: a lecture on the water cycle and river basins, drawing up a mental map of the Alcântara River and carrying out fieldwork on the banks of the Alcântara River channel, which runs just a few metres from the school. Finally, the students collected data and information on the perception of residents and shopkeepers about the Alcântara River by applying a questionnaire. Table 2 shows the contents and objectives of this module:

Table 2- Contents and Objectives of Module II

Contents	Water cycle Alcântara River Basin
Objectives	Review the Water Cycle and the dynamics of how a River Basin works
	analyse and interpret the students' socio-environmental perception of the Alcântara River using a mind map
	correlate the physical and environmental aspects responsible for the environmental degradation of the river Alcântara

Source: Prepared by the author Sousa (2017).

3.1.1.3 Module 3: Socio-environmental perception of the Alcântara River in São Gonçalo/RJ

In Module 3, the students drew up a map using ArcGIS *Online* with the aim of proposing solutions to the São Gonçalo/RJ City Council for the socio-environmental degradation of the Alcântara River from one of its channels. The students used data and information collected in the field and also gathered information through interviews with residents and shopkeepers in the neighbourhoods through which the river channel near the school runs (Figure 6).

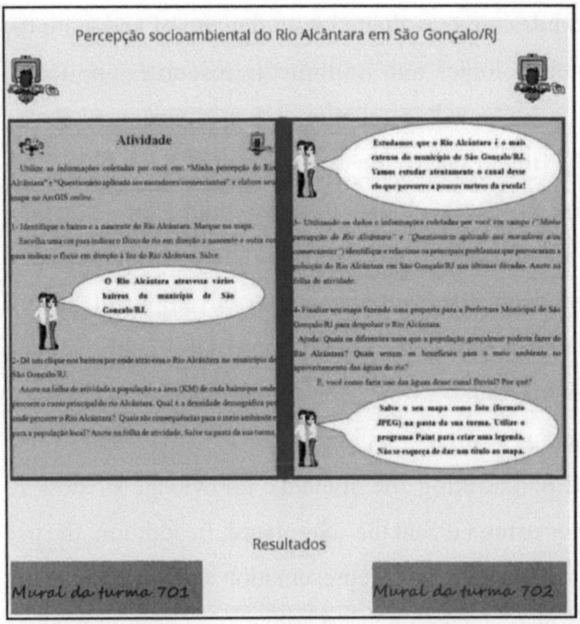

Figura 6 - MMR module 3 screenshots

Source: Prepared by the author Sousa (2014).

Table 3 shows the content and objectives of this module:

Table 3- Contents and Objectives of Module III

Contents	Basic elements of cartography River Basin Demographics Urban environmental problems Notions of Geoprocessing
Objectives	understand urban growth as one of the factors responsible for the degradation of the river channel;
	to identify on the maps produced in ArcGIS *Online* how the data and information collected by the students contributed to the socio-environmental perception of the Alcântara River from one of its channels;
	detect how the media (text, photo, video) have contributed to the socio-environmental perception of the Alcântara River;
	evaluate the contributions of using *Web* GIS to improve spatial perception;
	to evaluate the students' proposals to the São Gonçalo RJ City Council regarding the environmental degradation of the Alcântara River using basic cartography elements.

Source: Prepared by the author Sousa (2017).

3.2 Criteria used to evaluate Mapping My River

According to Hoffmann (2009), assessment is a process inherent to educational practices, conceived as progress, manifestations of students' doubts and difficulties in the development of activities.

Based on the evaluation discussed by Perrenoud (1999), characterised as formative observation, a continuous and integrated process judged day after day, the activities in the MMR modules were evaluated with the aim of verifying the level of contribution of geotechnologies and multimedia resources applied to cartography in the socio-environmental perception of the Alcântara River, without any concern for classifying or certifying the students.

The MMR evaluation process took into account the teaching methodologies used in geography classes, the relationship between the research teacher and the participating students, the level of cognitive development, as well as the school's infrastructure conditions and the organisation of this public school, including the entire management team.

In module 1 "Applying new digital technologies to cartography", the assessment process consisted of checking the students' knowledge of how to obtain geographical coordinates using GPS in the schoolyard. In addition, the use of legends and scales in drawing up a graphical representation of the home-school route using Google Earth was also checked. Observations were used to analyse students' attitudes towards geotechnologies and their contributions to cartography teaching.

With regard to module 2, "Exploring the Alcântara River in São Gonçalo/RJ", the aim was to diagnose the students' socio-environmental perception of the Alcântara River using a mind map. It should be added that the evaluation of this module was based on the students' participation, involvement and motivation in relation to the fieldwork around one of the channels of the Alcântara River, which runs a few metres from the school, as well as surveying the perception of residents and/or traders in relation to this river channel.

In the third module "Socio-environmental perception of the Alcântara River in São Gonçalo/RJ", the evaluation process was based on the graphic representation produced by the students with the help of ArcGIS Online. The proposal was to analyse the contributions of the Web GIS together with the geographical information collected in the field (interviews, surveys of natural and artificial objects present around the river channel, media recording - photo, text and video) in drawing up a proposal for the São Gonçalo/RJ city council, using a map, to minimise or resolve the degradation of the waters of the Alcântara River.

The results of this research were obtained by applying the MMR to two seventh-grade primary school classes; subsequently, the results between the classes were compared taking into account the following aspects:

a) the interest and behaviour of the students in taking part in the MMR. We sought to verify how geotechnologies and multimedia resources contributed to greater motivation and student participation in cartographic activities in Geography teaching in each of the classes taking part in the research.

b) analysing the contents of basic cartography and geographical knowledge. At this stage, we analysed how easy and/or difficult the classes found it to use cartographic and geographic knowledge in the activities carried out with Google Earth and GIS Web - ArcGIS Online.

c) students' performance in Geography during the period in which the tests were administered.

This stage verified the contributions of Mapping My River in relation to the students' bimonthly grades in Geography, particularly between the 2nd° and 4th° bimonthly grades.

Therefore, the evaluation of the MMR consisted of surveying the contributions, limits and possibilities of using geotechnologies and multimedia resources from a teaching methodology, as tools to aid the teaching of Cartography in Geography classes to develop the theme of socio-environmental perception.

CHAPTER 4

ANALYSIS OF MAPS AND MULTIMEDIA PRODUCED BY SCHOOLERS

Based on the action research methodology, this research considers the researcher and the students as subjects/objects in movement, in their action in the classroom, in an emancipatory process in the form of critical participation (GHEDIN E FRANCO, 2011), considering the teacher as a mediator - of meaningful geographical learning. We propose analysing each module of the course separately by class.

4.1 Analysing Module 1

This section analyses module 1 of the classes taking part in the MMR.

4.1.1 Analysis of Module 1: class 701

With regard to the use of GPS, we opted for the empirical method, taking into account involvement, participation and, above all, the identification of geographical coordinates (latitude and longitude).

This class, which was considered undisciplined and uninterested by the school's teachers, was very keen and enthusiastic to use the GPS in the schoolyard. We observed that the vast majority of the students had some idea of geographical coordinates, although there were difficulties in identifying the function and meaning of these basic elements of cartography through spatial orientation on the GPS. Difficulties and doubts were identified regarding the meaning and application of geographical coordinates, as the students were unable to identify them on the GPS which coordinate referred to latitude and longitude. Figure 7 shows the students using the GPS in the schoolyard.

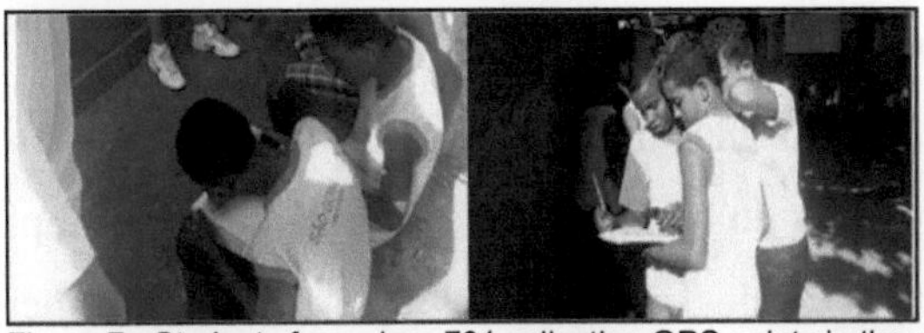

Figure 7 - Students from class 701 collecting GPS points in the school yard
Source: Personal archive (2013).

The activity in this module consisted of five steps in which the students created a spatial representation of the route from their home to school using Google Earth.

In the "First step", the students were asked to locate the school where they study and the house where they live by typing in a pair of geographical coordinates collected on the GPS under "Search" in Google Earth. Due to their difficulties with geographical coordinates, the students found it easier to type in the name of the school. With regard to the location of their home, many students didn't know the name of the street where they live or even any reference point in relation to their house.

Still on **the subject of "Step 1"**, the aim was to compare the visualisation scale of the satellite images, with the first representing the visualisation scale of the school and the second referring to the student's house. The aim was to recognise whether the student would apply the concept of scale in the geographical sense to explain the difference between viewing the school and the house in the images. The most similar answers were selected as "R= yes because it's very close to the school!", "the perimeter between my house and the school was big, I could only see it if I reduced the perimeter or reduced the map near my house and the school", "because there's a lot of bush and you can't see my house". We can see the students' difficulty with scale when they mention "perimeter" or even "shrinking the map".

Only one student carried out the **"second step".**

In **"<u>Step 3</u>",** the students were more involved, as all they had to do was click on the timer and they would be able to visualise the changes in the landscape of the place where they live. Two of the responses are worth highlighting:

a) "Ten years ago our street had no tarmac and the petrol station was closed. The community behind our houses was no smaller; there was no grocery shop, there was no car wash which is now a car dealership, there was no nursing home, there was no Raul Veiga pharmacy, there was no church near the school, there was no thrift shop, there was no oceanic church, there was no newsagent, there was no bus stop, there was no construction site, there was no flooring shop, there was no motorbike mechanic and there was no marble shop and no bakery!";

b) "They haven't tarmacked the streets, few houses have been built, this has only harmed the street there are lots of holes, lots of mud, the rubbish collector doesn't get through the dogs pee on the gate there are rats running over the street.", "tarmacking the streets the houses have changed they've cut down lots of trees they're going to build more houses buildings"

The answers above show the difficulties in perceiving the changes in the landscape of the neighbourhood where the students live, as there has been no correlation in its form, function or structure over the last ten years. Furthermore, there was no concern to portray the disorderly process of urban occupation in the neighbourhood.

The "<u>Fourth Step"</u> was interpreted and analysed together with the **<u>"Fifth Step"</u>;** the methodology included the spatial representation of the route from home to school using satellite images from Google Earth; they then used Paint to insert the title and legend. Figure 8 shows two maps that were drawn up on Google Earth.

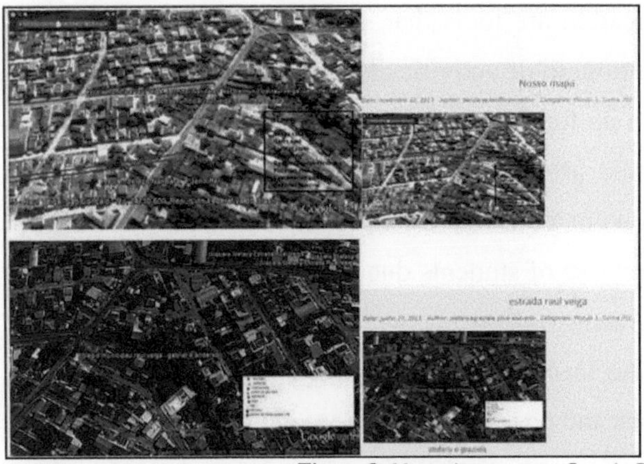

Figure 8- Maps drawn up on Google Earth
Source: Prepared by the author Sousa (2014).

The maps above showed an unsatisfactory result from the students in terms of using the basic elements of cartography, especially the legend, which was not related to the use of Google Earth or the Paint programme, but there were difficulties in terms of notions of vertical vision in order to read and interpret the spatial objects present in a satellite image.

We noticed that the cartographic categories of localisation, correlation and analysis were not developed in previous school years. As a result, there is no correlation between the spatial objects represented in the legend whose purpose was to create a map to represent the route from home to school.

The captions were confusing; firstly, students Nandara and Jenniffer Veríssimo selected the colours randomly without correlating the natural and artificial elements on the map; secondly, in the caption prepared by students Stefany and Graziela, the use of colours did not show effective communication between the colours and the objects represented.

It is important to emphasise that this activity was not designed to develop the basic elements of cartography, nor the concepts of place and landscape. It was assumed that the students would have learnt these concepts in the 6th grade° and in previous school years.

We found that the use of digital technologies applied to cartography, GPS and the Google Earth application are tools that stimulate and arouse students' interest in participating in cartographic activities.

The lesson plan drawn up for this module was organised into ten lessons, but the activity was finalised after almost three months due to the limited availability of computers, eight in total, and low Internet connection and network distribution, as well as the release of students due to the absence of teachers on the days of the MMR assessment.

In order to improve the teaching of cartography, it is necessary to take advantage of the contributions and possibilities of the technologies applied to the mapping process, so that the limits do not become obstacles to the use of these tools as an aid to the teaching of cartography.

4.1.2 Analysis of Module 1: class 702

Class 702 was very participative and interested. The students were enthusiastic. They all wanted to go at the same time, but there were only two GPS devices available. Figure 9 shows a group of students using the GPS in the schoolyard.

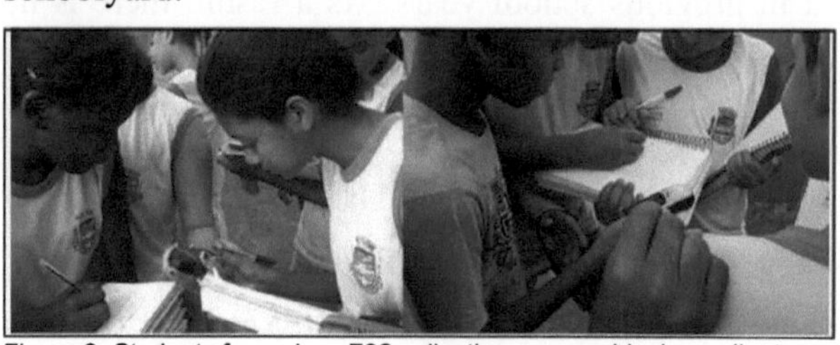

Figure 9- Students from class 702 collecting geographical coordinates on the GPS

Source: Prepared by the author Sousa (2014).

The use of GPS was analysed through observations of involvement, participation and, above all, the recognition of geographical coordinates (latitude and longitude).

Despite the students' interest in the functions of the GPS and the collection of latitude and longitude points in the schoolyard without mathematical calculations, many students had difficulties in knowing the geographical coordinates for locating a point on the earth's surface. Many students inverted the identification of latitude and longitude on the GPS, as they were unable to differentiate between the geographical coordinates.

As well as collecting geographical coordinates and recording photos, this activity comprised five steps that resulted in the creation of a map to represent the route from home to school using Google Earth.

Due to the difficulties in identifying the geographical coordinates on the GPS in the "First step", the students located the school by typing its name into the "Search" toolbar; the house was located by its street address.

Still in "Step 1", the comparison of the satellite images, the first of which represented the school and the second the student's home, sought to apply the concept of scale, in the geographical sense, to explain the difference in the visualisation of these two spatial objects in the images. The most similar answers were selected:

a) "not because it was too far away and I had to zoom in to find it."
b) "no, because it's too far away and I can't tell exactly how far it is from my house."

We noticed that the students used the term distance as a measure to explain the visualisation of the objects (house and school) in the two satellite images.

The "Second step" was not analysed because no student recorded a photo of the route from home to school.

Completing **"Step 3"** was stimulating for the students, as they were able to visualise the changes in the landscape of the place where they live over the last ten years in just one "c/ick". The answers showed the process of building the neighbourhood, but without any concern for the natural aspects, showing little or almost no relationship with nature, know:

a) "A lot of things have changed, especially my house, my street, more shops have been built, houses have been remodelled, streets have been paved";

b)"The streets used to be like this, they weren't asphalted, there were no spring breakers, the streets were very muddy, there was no way to walk, the residents didn't know what to do because the town hall didn't want to asphalt the streets, at that time the houses were all made of bricks. Time went by and some things changed, like the streets were tarmacked and the houses changed for the better and the town hall helped a lot, but we need a lot more. The residents aren't complaining any more, but at that time there was a lack of water, electricity etc... Now everything is different, there's water. There's electricity and there's a lot more. The streets are tarmacked, there are lots of spring breakers, etc."

The "Fourth step" was interpreted and analysed together with the **"Fifth step";** the proposal was to spatially represent the home-school route using satellite images from Google Earth. They used the Paint programme to insert the basic elements of a map. Figure 10 shows the maps that were created in Google Earth.

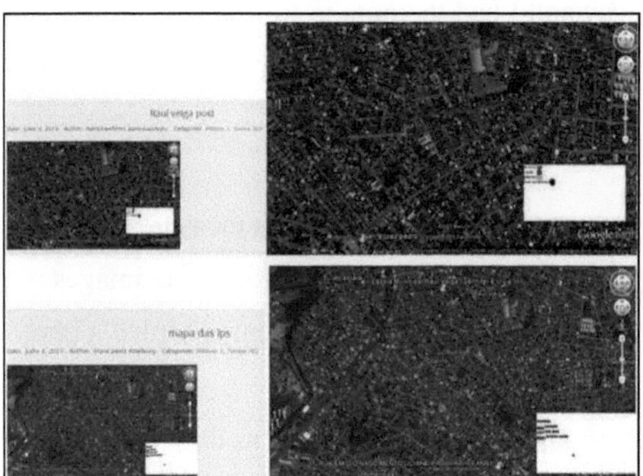

Figure 10 - Maps created in Google Earth

Source: Personal archive (2014).

We found that the map legend was constructed using the variable visual colour and there is a correlation between the elements represented graphically.

The students' participation, involvement and interest throughout the activity showed how GPS and the use of the Google Earth application are tools that stimulate and arouse students' interest and, consequently, become important resources for teaching maps to school-age teenagers who are contemporary with digital technologies.

The use of geotechnologies such as remote sensing through Google Earth and GPS must take into account basic knowledge of cartography as well as geographical knowledge when producing a digital map.

4.2 Analysing Module 2: Exploring the Alcântara River in São Gonçalo/RJ

This section analyses Module 2 of the classes taking part in the MMR.

4.2.1 Analysis of Module 2: class 701

Firstly, we wanted to get to know the students' perception of the Alcântara River from one of its channels, which runs just a few metres from the school. The PCNs for Geography (BRASIL, 1998) describe place as an important concept for understanding Geography as a discipline that unravels the nature of places and the world as a habitat for man.

The analysis and interpretation of the mind maps was based on Kozel (2007), as well as Teixeira and Nogueira (1999).

In order to discuss the importance of mind maps in understanding students' socio-environmental perception of the Alcântara River, four mind maps were selected, divided into two groups, according to the specifics below:

4.2.1.1 mental map and nature "untouched" by man;

4.2.1.2 mental map and the proximity of the socio-environmental "reality" of the Alcântara River canal;

These spatial representations were analysed and interpreted separately by class.

4.2.1.3 Analysis of Module 2 regarding mind maps: class 701

In the first stage of this module, the students were asked to draw up a mind map based on the drawings made from their experience with the Alcântara River. Four mind maps were selected, interpreted and analysed according to the specifics of item 4.2.1.

The maps in Figure 11 present a feeling of idealised nature (Rio with its crystal-clear waters, trees, air without pollutants).

Figure 11- Mind maps and nature "untouched" by man - class 701

Source: Prepared by the author Sousa (2014).

The imaginary of these students is formed by a mythical conception of nature, as if the area around the Alcântara River were a protected natural area, a paradise, an uninhabited space in which nature remains virgin and untouched by man, according to Diegues (2001). We see a representation of preservationist ecologism that doesn't show the artificial and natural objects that make up this geographical space. In accordance with Santos (2002, p. 153), "Space should be considered as a set of relationships realised through functions and forms that bear witness to a history written by past and present processes".

Therefore, there is no geographical reading of the surroundings of the Alcântara River Canal represented on these students' mental maps. In addition, no legend was found, which also made it difficult to read these maps.

We found the closest representation of the Alcântara River channel with its polluted waters, as shown in the mental maps in Figure 12.

Figure 12 - Mental maps and proximity to the socio-environmental "reality" of the Alcântara River canal - class 701

Source: Prepared by the author Sousa (2014).

In both mental maps, the factors responsible for the canal's degradation were not identified. In the mind map on the left, there was the notion of a vertical view of the banks of the canal with a spatial reduction, but an uninhabited and lifeless space, and there was also a concern to legitimise the power of the authorities to prevent attitudes and actions that pollute the river channel; while in the mind map on the right, there was also a vertical view of the canal and a frontal view of its banks, but the spatial objects were not connected to the represented space.

When analysing and interpreting the mind maps of class 701, we detected that the students were in agreement about the pollution of the waters of the Alcântara River channel, as the spatial representations did not, for the most part, show prospects for programmes and actions to clean up this water resource; however, they did identify the degradation of the waters of this river, but without correlating the processes that have contributed to the current socio-environmental situation of this water resource.

4.2.1.4 Analysis of Module 2 regarding mind maps: class 702

Four of the class's mind maps were selected to be analysed and interpreted, according to the specificities outlined in the item

Within this specificity, the mental maps represent nature without the action of man. Figure 13 shows mental maps and nature "untouched" by man.

Figure 13 - Mind maps and nature "untouched" by man - class 702

Source: Prepared by the author Sousa (2014).

When analysing the students' mental maps shown in figure 40, it can be seen that there are no natural objects or artificial objects around the Alcântara River channel, as if the river channel were dehumanised, there were no buildings, in other words, a representation of a space without the presence of man. There is a notion of naturalist myth which, according to Diegues (2001, p.32), "[...] refers to a symbolic representation in which there are natural areas untouched and untouchable by man".

Archela, Gratão and Trostdorf (2004) add that mental maps make it possible to establish relationships between the way each person sees their place. In this sense, the students perceive the Alcântara River canal as a clear water resource with life and a naturalistic charge, since it did not represent human elements and the landscape built on the banks of the canal, i.e. there is no conception of geographical space.

With regard to the mental maps that are closest to the socio-environmental "reality" of the Alcântara River canal, the spatial representations shown in Figure 14 show the selected mental maps that are closest to reality.

Figure 14- Mental maps and proximity to the socio-environmental "reality" of the Alcântara River channel - class 702

Source: Prepared by the author Sousa (2014).

According to Richter (2010, p. 167), "[...] the map is not a direct reflection of reality, it undergoes filters, particular readings that alter its dimension - from the general to the specific".

In this sense, analysing the mental maps in Figure 14 shows the perception of the river channel as it "really is", with its dark waters due to the objects thrown along its course; however, there was a limitation in representing the pollution of the channel's waters, ignoring the riverbanks, the spatial objects in the surroundings, such as houses, shops and workshops, without relating them as agents that produce urban space and are responsible for the current environmental situation of this channel.

In the mind maps of class 702, we found that the vast majority of students did not represent correlations between pollution and the natural and environmental factors responsible for the degradation of the canal's water, which has made it unfit for use and consumption.

Therefore, in both classes we saw some difficulties in terms of cartographic notions, such as proportionality between the canal represented or, even the main course of the Alcântara River and the spatial objects around this water resource, in relation to the orientation of spatial objects, such as homes, shops, streets which, in many spatial representations were placed without correlation with the surroundings of this canal; however, the waters of the canal or the main course of the river were represented as the most significant objects on the map.

4.2.1.3 Analysis of "My socio-environmental perception of the Alcântara River"

In this stage, the students went into the field to gather information on the physical-environmental and social aspects of the Alcântara River channel, located a few metres from the school. Each pair or trio was given a form on which they could record their perception of the Alcântara River channel under study. The analyses were done separately by class.

4.2.1.4 Analysis of "My socio-environmental perception of the Alcântara River": class 701

With regard to class 701, the results of the survey are shown in Figure 15.

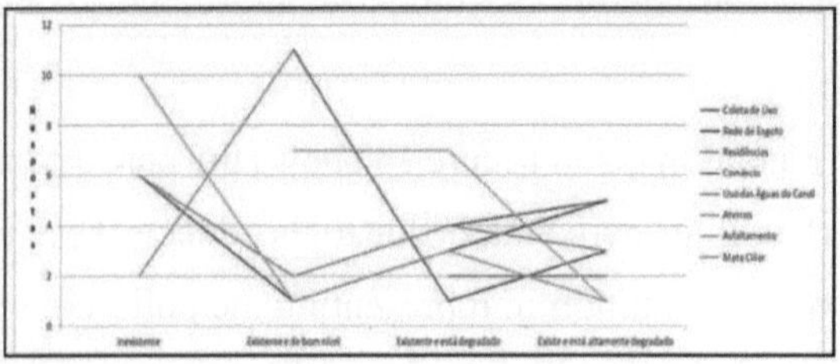

Figure 15 - Socio-environmental perception of students in class 701 in relation to the river channel of the Alcântara River, which runs close to the school.
Alcântara that runs close to the school
Source: Prepared by the author Sousa (2014).

When analysing Figure 15, the students classified the waste collection system and the sewage system as existing around the canal, but at a high level of degradation. According to these students, the precariousness of the sewage treatment system, the deficiency of rubbish collection around the canal and the absence of riparian forest are factors that have contributed to silting, making the waters of the canal unfit for consumption.

The tarmac and embankments were classified as having a medium level of degradation.

According to the students' level of perception, buildings such as residences and businesses exist along the Alcântara River channel with a low level of degradation, which means that the students may not have considered the basic infrastructure of these buildings.

4.2.1.5 Analysis of "My socio-environmental perception of the Alcântara River": class 702

When analysing the results of the survey on the physical, environmental and social aspects of the canal's surroundings, class 702 obtained the results shown in Figure 16.

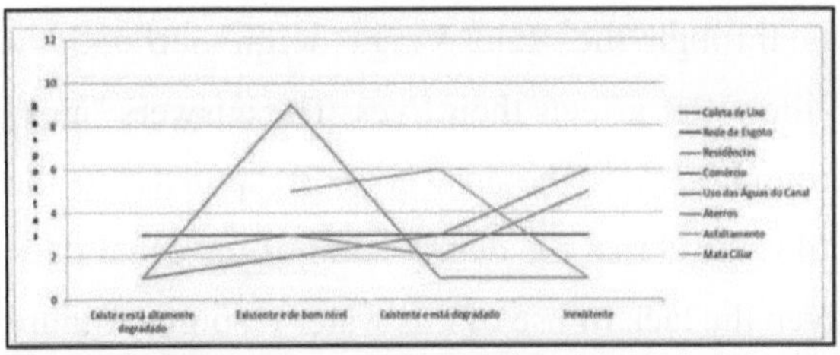

Figure 16- Socio-environmental perception of the students in class 702 in relation to the river channel of the Alcântara River, which runs close to the school.
Alcântara that runs close to the school

Source: Prepared by the author Sousa (2014).

The responses detected the rubbish collection system and the riparian forest at a high level of degradation. Within the medium level of degradation around the canal were found the asphalting and landfills.

The students classified the landfill and the riparian forest as non-existent around the canal, but there was some confusion when they classified the riparian forest as "existing and highly degraded".

The students therefore realised that the lack of rubbish collection around the canal and the absence of riparian forest are factors that have contributed to the degradation of the river canal that runs just a few metres from the school.

4.2.1.6 Analysis of the students' socio-environmental perception

In order to obtain information and collect data on the Alcântara River channel, which runs just a few metres from the school, the students interviewed the residents and/or shopkeepers with the aim of understanding their socio-environmental perception of this water resource. From the responses, the questions corresponding to numbers 6 and 7 were selected.

4.2.1.7 Analysis of the questionnaire applied by the students to the residents and/or shopkeepers in class 701

Twenty-one residents and/or shopkeepers were interviewed about the

Alcântara River canal that runs through the Raul Veiga neighbourhood, in particular whether the quality of the water affects their lives. The answers "Yes" and "No" were very divided; around 57% of the answers corresponded to the polluted waters of the canal and 43% said that the degradation of the canal does not influence their daily lives, given the lack of use of this water resource. Figure 17 shows the perception of the waters of the Alcântara River canal among the residents and/or shopkeepers interviewed by the students.

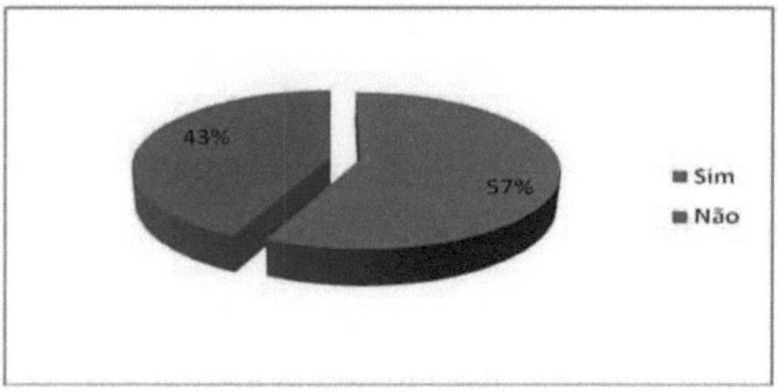

Figure 17 - Does the quality of the water in the Alcântara River channel affect your life?
Source: Prepared by the author Sousa (2014).

Almost all of the interviewees, with the exception of 7%, perceive the river channel as a ditch due to the marked degradation and pollution of its waters. Figure 18 shows the perception of residents and/or traders regarding

on the quality of the waters of the Alcântara River from one of its channels.

Figure 18 - How do you see the quality of the water in this river?
Source: Prepared by the author Sousa (2014).

4.2.1.8 Analysis of the questionnaire applied by the students to the residents and/or shopkeepers in class 702

With regard to the perception of residents and/or shopkeepers about the Alcântara River canal that runs near the school, it was found that 67% of those interviewed said that the polluted waters of the canal affect their quality of life, while a minority of them, around 33%, mentioned that the degradation of the canal does not influence their daily lives because they do not use the waters. Figure 19 shows how the quality of the water in the Alcântara River canal affects those interviewed.

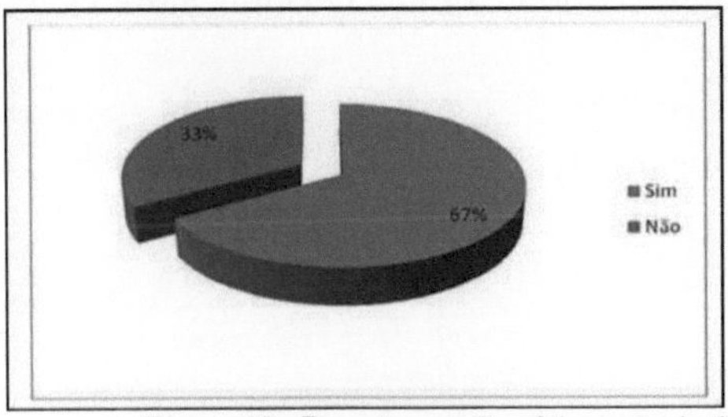

Figure 19 - Does the quality of the water in the Alcântara River channel affect your life?

Source: Prepared by the author Sousa (2014).

With regard to the quality of the water in this river, almost all of the interviewees, with the exception of 8%, perceive the river channel as a ditch, as its waters are at a high stage of degradation. Figure 20 shows the perception of residents and traders regarding the quality of the water of the Alcântara River from one of its channels.

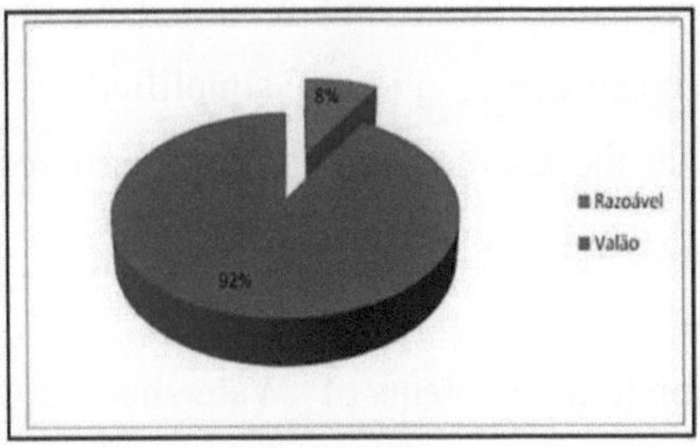

Figure 20 - How do you see the quality of the water in this river?
Source: Prepared by the author Sousa (2014).

4.3 Analysing Module 3: Socio-environmental perception of the Alcântara River in São Gonçalo/RJ

Module 3 was assessed on the basis of information collected in the field about the Alcântara River channel, which runs just a few metres from the school, through observations, interviews with residents/business owners and media recording (text, photo and video). In item 4 of this activity, the students produced maps in ArcGIS Online with proposals and solutions for the São Gonçalo/RJ City Council to clean up the Alcântara River.

Again, the classes were assessed separately.

4.3.1 Analysis of Module 3: class 701

In the first stage of the activity designed for this module, the students had to identify the source and mouth of the Alcântara River using the "Content" that corresponds to the layer in a GIS, "São Gonçalo contour lines" and the *"OeenSreeetMap"* Base Map (available on ArcGIS *Online*). We asked the students to choose a symbol from the features (symbols) and colours available in this Web GIS to mark the source and mouth of this river.

The students had difficulty identifying the source and mouth from the contour lines, so the researcher reminded the students, in a very simplified way so as not to interfere with their answers, that the source is where the river rises and the mouth is where the river meets the sea, the Alcântara River being a tributary of the Guaxindiba River.

As a result of the difficulties in relation to the contents of "Watershed" and "Contour line", the students identified the source and mouth of the Alcântara River on the map by "following" the name of this water resource and, with the marker using the point deployment mode and the colour and shape variable, indicated the flow towards the source and mouth of the Alcântara River. and consulted the website http://www.mapeandomeusrios.com.br/modulo-2/ where some socio-environmental information about the Alcântara River is briefly available. Figure 21 shows the identification of the source and mouth of the Alcântara River.

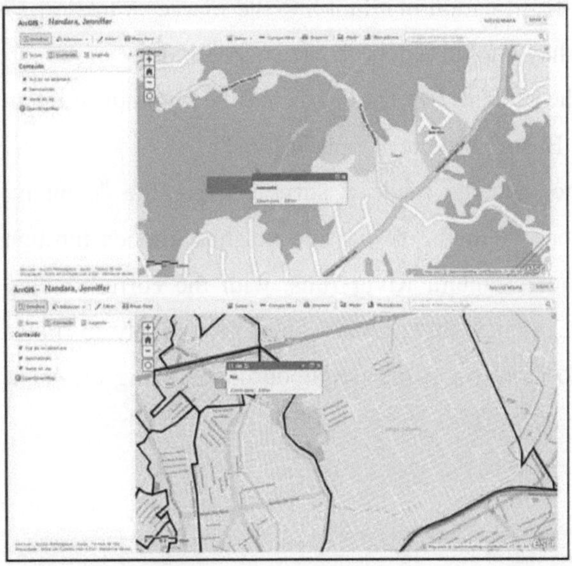

Figure 21 - Identification of the source and mouth of the Alcântara River on the *OpenStreetMap"* Base Map
Source: Prepared by the author Sousa (2014)

We can see in Figure 21 that identifying the source and mouth of the river proved to be "strange" content in the graphical representation, content that should have been learnt by the students in the sixth year of schooling. The answers obtained by the students can be analysed in Figure 22.

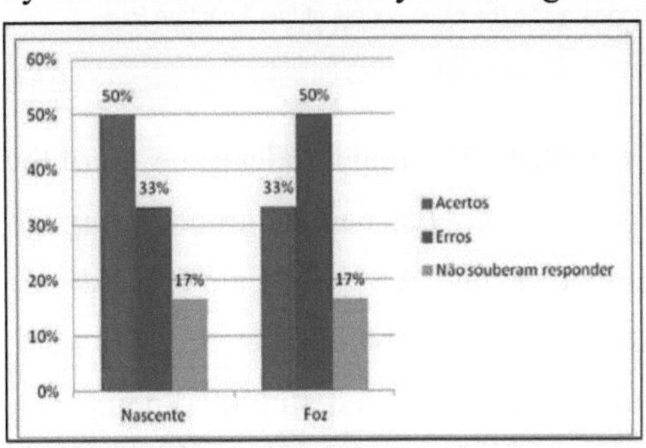

Figure 22- Percentage of identification of the source and mouth of the Alcântara River

Source: Prepared by the author Sousa (2014)

The markers in the direction of the mouth indicated a higher number of correct answers (50 per cent), although this corresponded to half of the students taking part, compared to the identification in relation to the source. A total of 17 per cent of the students said they couldn't identify the source and the mouth. In this sense, it can be seen that the students have not absorbed the minimum content relating to the "Hydrographic Basin" and the "Contour Line".

In the second step, the students made a survey, using the "Content: Alcântara River Neighbourhoods", of the respective neighbourhoods through which the main course of the Alcântara River runs, with the corresponding population and Km^2 area in the municipality of São Gonçalo/RJ. Figure 23 shows the correct answers related to identifying the neighbourhoods along the main course of the Alcântara River.

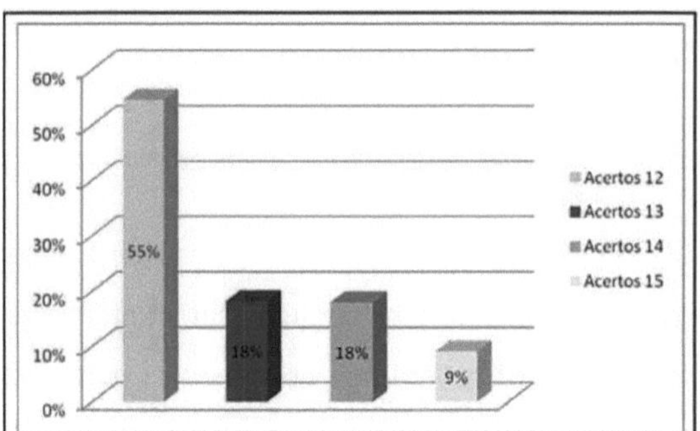

Figure 23 - Identification of the neighbourhoods along the main course of the Alcântara River
Source: Prepared by the author Sousa (2014).

Looking at Figure 23, it can be seen that 55% of the students, i.e. 13 of the 24 participating students, located and analysed the neighbourhoods through which the main course of the Alcântara River runs in São Gonçalo/RJ, and only 18%, i.e. only 4 students correctly indicated the neighbourhoods.

These results clearly show that the students, as Simielli (2003) points out, have not left the first level of cartographic category, which is localisation and analysis, and he adds that 7th graders are already able to work with analysis/localisation of spatial phenomena represented on the map because they have gone through the cartographic literacy process; therefore, these categories have not been grasped by these students.

It is necessary for these students to return to cartographic literacy, in other words, to build notions of the basic elements of cartography so that students can understand the spatialisation of geographical phenomena through the preparation of maps and graphs, considering the process of teaching and learning cartography in geography education as a language, as stated by Passini (2012).

The students produced a text listing the reasons and human actions that have contributed to the degradation of the Alcântara River over the last few decades, based on the information collected in the field: "My Perception of the Alcântara River" and "Questionnaire applied to residents/traders"; they were also asked to record media (photo and video) of the canal's surroundings.

Due to the time it took to complete the research, which coincided with the end of the school year, the text was written as a homework assignment, rather than in the school's computer lab; photos and videos were also taken outside of school hours without the researcher's supervision.

The main factors responsible for the pollution of the channel, described in the texts produced by the students, were summarised in a graph for better interpretation of the answers. Figure 51 shows the elements identified as responsible for the pollution of the Alcântara River, from one of its channels.

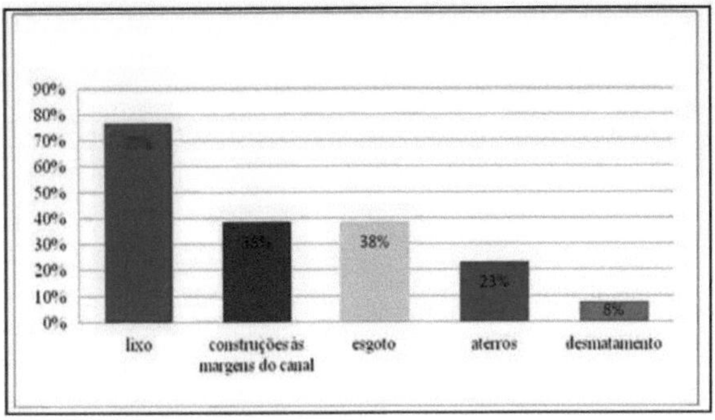

Figure 24- Sources of pollution in the Alcântara river channel near the school

Source: Prepared by the author Sousa (2014).

As can be seen in figure 24, the students' perception of the sources of pollution of the Alcântara River channel, located near the school, is related to the following anthropogenic factors: lack of rubbish collection, responsible for 77 per cent of the change in the channel's degradation, followed by sewage disposal with 38 per cent and constructions (houses, shops and mechanical workshops) on the banks of the channel, which also contributed 38 per cent to the degradation of its waters:

> **"[...] the growth of urban areas without the necessary conditions for** the maintenance of green areas, to allow the balance of the hydrological cycle, without the minimum sanitation conditions (rubbish, sediments and sewage), are examples of indirect impacts, originating in **the drainage basin and causing the degradation of the channels."** (CUNHA, 2007, p.224)

Two videos and a photo were produced showing their perception of the causes of pollution of the river channel; the videos are available at the following website: http:< //www.mapeandomeusrios.com.br/mural-turma-701/>. Figure 52 shows the photo and images of the videos produced by the students using their smartphones.

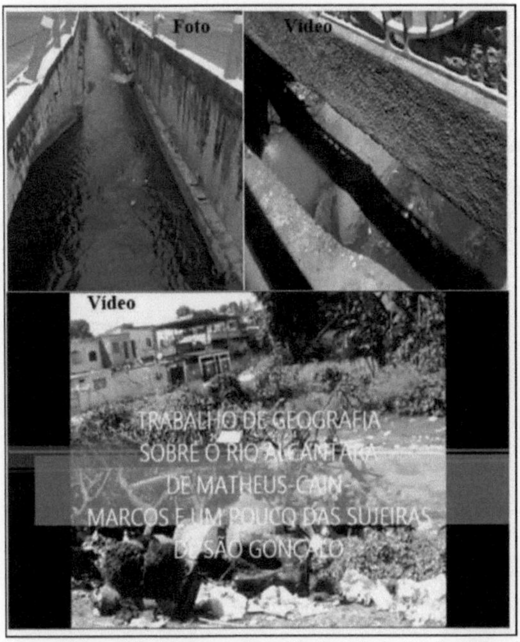

Figure 25 - Media produced by the students about the Alcântara River channel near the school. Source: Prepared by the author Sousa (2014).

The multimedia resources constructed by the students (photos and videos), although they represent a small proportion, shown in Figure 25, portray how residents and shopkeepers in the Raul Veiga neighbourhood, as well as the students, have realised that sewage discharged directly into the canal without

treatment and inefficient rubbish collection are the main causes of the degradation of this canal, whose banks are undermined and eroded.

The video "Geography work on the Alcântara River by Matheus- Cain-Marcos and a bit of the filth of São Gonçalo/RJ" points out the lack of enforcement policies regarding irregular occupation of the banks of the Alcântara River channels, rubbish collection and sewage treatment, as well as the lack of affective ties between people and the Alcântara River channels due to the dumping of objects such as furniture and sofas; while the other video produced only showed the environmental situation of the river. There were no proposals for minimising and finding solutions to the degradation of the river's waters.

In the last stage of this module, the students made a proposal in an attempt to resolve the environmental degradation of the Alcântara River from one of its channels by means of a map using ArcGIS Online. The aim was to assess how the students used their knowledge of cartography in drawing up the proposals, proposing solutions to minimise or even improve the socio-environmental conditions of the Alcântara River from one of its channels that runs close to the school, as well as their geographical knowledge in producing the map. Figure 26 shows the solutions proposed by the students to clean up the river channel.

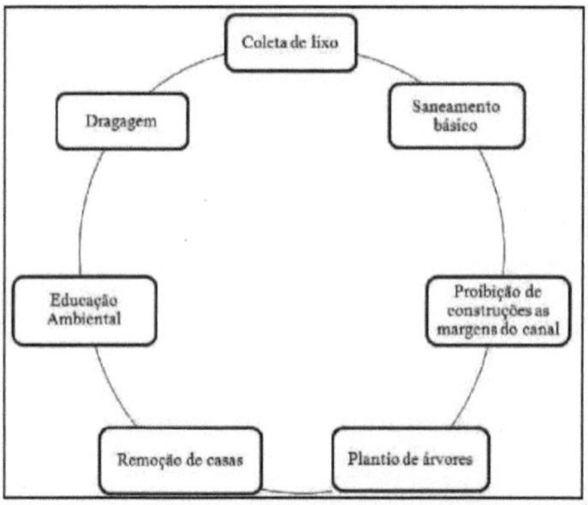

Figure 26 - Solutions proposed by the students to clean up the river channel

Source: Prepared by the author Sousa (2014).

We noticed that the students realised how the canal's socio-environmental conditions had been affected by anthropogenic actions and consequently proposed a set of compensatory mitigation measures that could solve or even reduce the intensity of environmental degradation from the direct impacts of polluting sources.

However, when analysing the knowledge of cartography that the students had supposedly learnt in previous years of schooling, it emerged that they arrived at seventh grade with difficulties in understanding the vertical view from satellite images of their living space.

When analysing the map, there were difficulties in relation to the essential element in cartography, which is the legend, which, according to Martinelli (2003, p.34) "is a guide to reading the map. In a first contact, the legend has the role of relating all the signs used on the map, indicating what they mean", according to the intentions and the historical and social context of those who drew it up. Figure 27 shows a map produced by two students with difficulties in cartographic literacy.

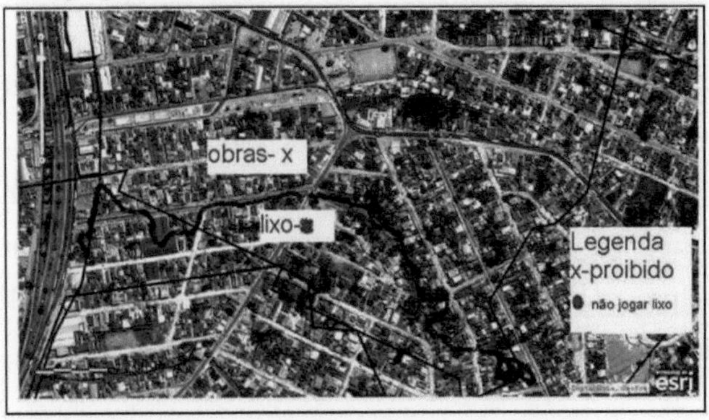

Figure 27- Analysis and interpretation to clean up the Alcântara River channel
Source: Prepared by the author Sousa (2014).

Of all the maps produced by the students, only two were able to partially decode the satellite image; the students constructed a map using the visual variables colour and shape, proposing that the City Council place rubbish bins on public roads and remove nearby houses on the banks of the Alcântara River canal, which is located a few metres from the school in the Raul Veiga neighbourhood; However, they had difficulties interpreting the size of the objects due to the distortion of the size of the bins, overtaking the bins, overtaking the houses and even placing them in the bed of the canal. Figure 28 shows a proposal to minimise or even clean up the river channel.

Figure 28 - Analysis and interpretation to clean up the Alcântara River channel

Source: Prepared by the author Sousa (2014).

How did these students go through "cartographic literacy", especially in the 6th grade° , where Cartography is a thematic axis of the Geography PCN for the third cycle, and fail to propose actions to recover the canal's waters through cartographic representation?

4.3.2 Analysis of Module 3: class 702

In the first activity of this module, the students had to identify the flow of water towards the source and mouth of the Alcântara River using the "Content" layer in a GIS, "São Gonçalo contour lines" and the *"OeesSreeetMap"* Base Map, using the point shape deployment mode and choosing the visual variables colour from the symbols available in ArcGIS *Oslise*. Figure 29 shows an example of the identification towards the source and mouth of the Alcântara River.

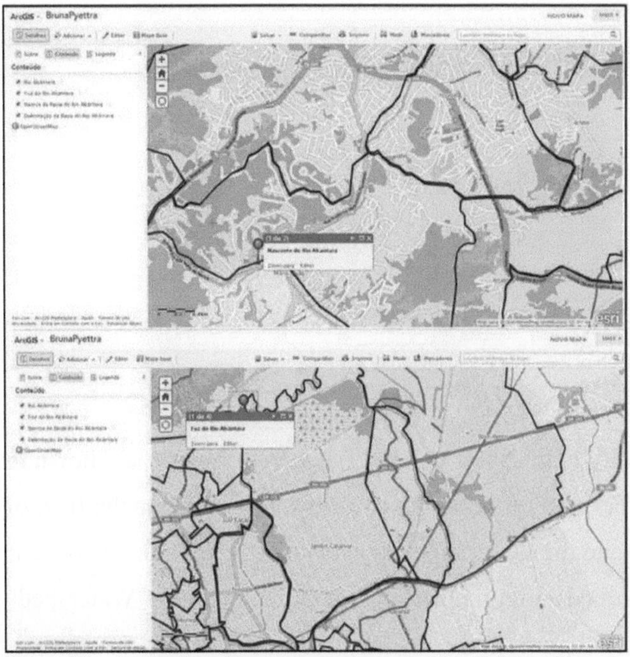

Figure 29 - Identification of the source and mouth of the Alcântara River on the *OpesSteeetMap* Base Map
Source: Prepared by the author Sousa (2014).

We noticed that this class had difficulties locating the source and mouth using contour lines, even though the proposal for this activity took into account the students' previous knowledge of "River Basin" and "Contour Line" in the 6th grade° ; therefore, the source and mouth of the Alcântara River were identified by consulting geographical information about the Alcântara River in Module 2, which is available at http:<//www.mapeandomeusrios.com.br/modulo-2/>. Figure 57 shows the correct answers related to identifying the neighbourhoods along the main course of the Alcântara River.

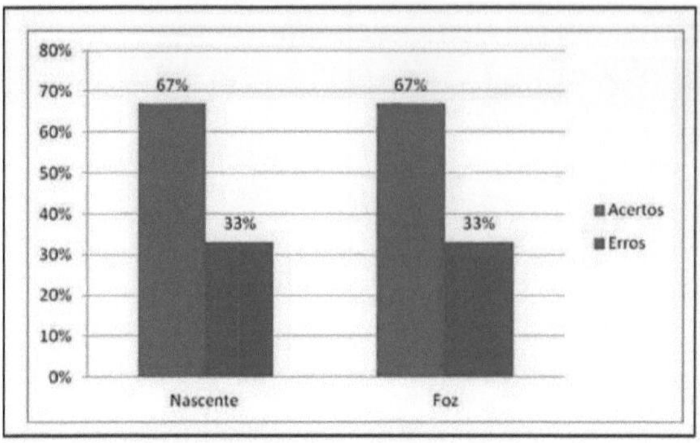

Figure 30 - Percentage identification of the source and mouth of the Alcântara River

Source: Prepared by the author Sousa (2014).

The level of difficulty in this class can be considered average, as shown in Figure 30, given that 50 per cent of the students were able to indicate the flow of water towards the source and the same percentage towards the mouth. It was clear that the students had absorbed a minimum of content relating to the "Watershed" and "Contour lines".

In the second stage of this activity, the students located, using the "Content: Alcântara River Neighbourhoods", the neighbourhoods through which the main course of the Alcântara River runs in the municipality of São Gonçalo/RJ, along with the population and the corresponding Km^2 area. Figure 31 shows the hits related to identifying the neighbourhoods along the main course of the Alcântara River.

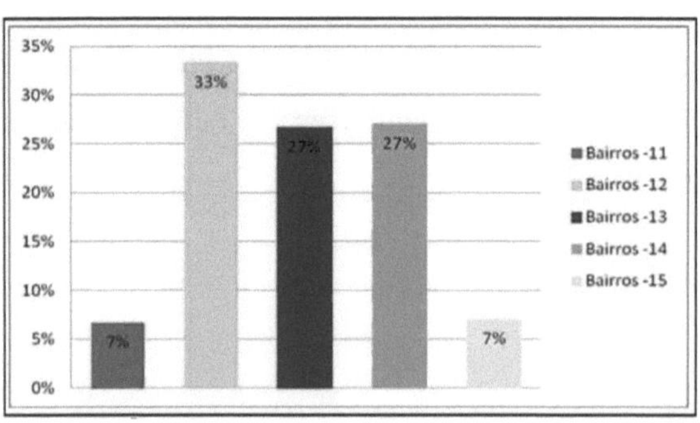

Figure 31- Identification of the neighbourhoods along the main course of the Alcântara River
Source: Prepared by the author Sousa (2014).

Looking at Figure 31, we can see that 12 neighbourhoods are identified where the main course of the Alcântara River runs in São Gonçalo/RJ, although the answers were close to 14 neighbourhoods, which corresponds to the total number of neighbourhoods that cross the main course of the Alcântara River.

The cartography taught in schools seems to be unsatisfactory, as the students had difficulties with localisation/analysis, which is the most basic level of map teaching in the classroom. This activity was aimed at 7th graders° who, according to Simielli (2003, p. 95) "[...] will already be able to work with analysis/localisation, with correlation". We need to return to cartographic literacy so that these students are able to encode and decode the geographical information represented on the map.

The teaching of maps in schools should point out the locations of geographical facts and phenomena, but it cannot be restricted to this first cartographic category, as (SIMIELLI, 2003) points out; students need to be able to analyse and correlate the spatial information represented. In this sense, the teaching of maps should be worked on as a language of graphic communication of geographical space.

> The map, therefore, is of the utmost importance so that anyone who is interested in more rational travel, in understanding the distribution and organisation of spaces, can inform themselves and be informed.
>
> (ALMEIDA and PASSINI, 2005, p. 16).

In item 3 of this activity, the students produced a text listing the reasons and human actions that have contributed to the degradation of the Alcântara River, based on the following information collected in the field: "My Perception of the Alcântara River" and "Questionnaire applied to residents/business owners". It is important to emphasise that the text would initially be produced digitally on the Internet, at the following address: http:<//www.mapeandomeusrios.com.br/atividade-2-modulo-2/, However, due to the poor Internet connection, the students wrote the text as a homework activity.

The students' survey of the sources of pollution in the Alcântara River channel near the school was put together in a graph for better interpretation. Figure 32 shows the students' perception of the factors responsible for the degradation of the Alcântara River.

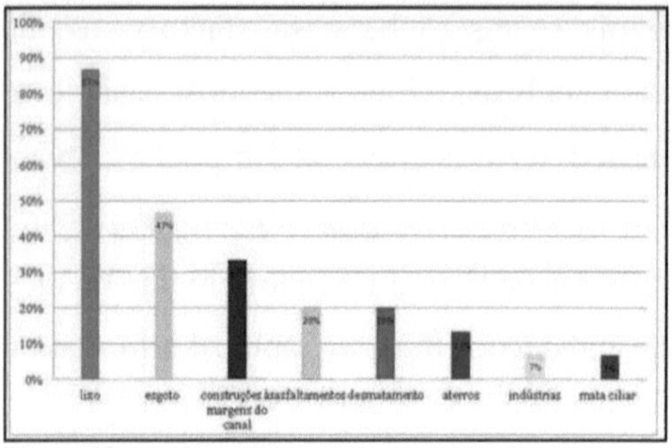

Figure 32 - Sources of pollution in the river channel near the school

Source: Prepared by the author Sousa (2014)

The students realised that the most significant factor in the degradation of the Alcântara River channel located near the school is the lack of rubbish collection, which is responsible for 87% of the change in the channel's degradation, and secondly, the lack of rubbish collection, which is responsible for 87% of the change in the channel's degradation.

place, sewage is discharged directly into the canal without treatment, which in turn is responsible for 47 per cent of the impairment of its water quality.

In addition to the text, the students produced other media such as videos and photos of the canal's surroundings, which were used as tools to improve the perception of the spatial dynamism of the river canal's surroundings near the school. These multimedia are available at the following e-mail address: http:<//www.mapeandomeusrios.com.br/ mural-turma-702/>. Figure 33 shows the photo and images of the videos produced by the students using their smartphones.

Figure 33 - Media produced by class 702 students about the Alcântara River channel near the school

Source: Prepared by the author Sousa (2014).

The media productions were made by half the class. Through photos and videos, the students showed the degradation of the waters of this canal with its polluting sources and the consequences, among them: silting and flooding in the main course of the Alcântara River; however, no proposal or initiative was identified to reduce or even clean up the waters of this canal.

In the last activity item of this module, the students analysed and correlated all the information they had collected in the field, as well as their perceptions and experiences on the journey from home to school, in order to propose solutions to the problem of the environmental degradation of the river.

Alcântara from one of its canals. ArcGIS Online was used to create a spatial representation of this proposal. Figure 34 shows the solutions proposed by the students to clean up the river channel.

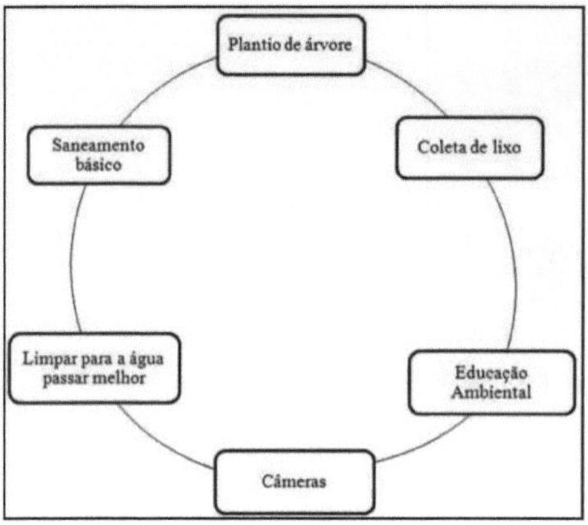

Figure 34- Solutions proposed by the students to clean up the river channel

Source: Prepared by the author Sousa (2014).

The proposals are in line with the perception of the students, who pointed to the lack of adequate rubbish collection and basic sanitation as the main contributors to pollution of the river channel. Therefore, the students showed that it is necessary to carry out, among other compensatory mitigation measures, rubbish collection and sewage treatment around the canal.

In agreement with Almeida and Passini (2005), the title of the map is important for knowing the spatial information represented graphically. However, when analysing the maps produced by this class, firstly, it was found that none of the students gave a title to their map, i.e. they were unable to read their own map.

The legend is the fundamental element in cartographic language that makes it possible to encode and decode spatial information represented by symbols and signs. The students in this class were unable to organise a system of signs and colours that would establish a system of

signs with their colours and shapes as a proposal for the degradation of the Alcântara River channel through spatial representation.

These basic elements of cartography should have been learnt by the students, especially in the 6th° grade, when cartography is considered content, but this was not seen in these maps.

° These students arrived at the 7th grade with difficulties in locating and interpreting, and also showed difficulties with notions of vertical vision, based on information extracted from the satellite image referring to the objects that make up their everyday space, in this case, the school's surroundings.

These map analyses showed that these students need to go through "cartographic literacy" again, as they were unable to use this cartographic knowledge to propose improvements or even actions to recover the quality of the canal's water using cartography.

4.3.3 Comparative analysis of results between classes 701 and 702 at Raul Veiga Municipal School

The comparative analysis of the MMR results between the classes took into account student participation, knowledge of cartography and geography and, lastly, school performance in geography during the course of the tests. The school's infrastructure and its impact on the research had to be taken into account.

We found that both classes showed an interest in taking part in the MMR, even at times outside of Geography lessons, whether at break time or even during lessons in other subjects.

With regard to Module 1, in class 701 there were 9 students who didn't take part in the activities, 4 of whom joined in the middle of the 1st semester° ; in class 702, the dropouts totalled 5 students. The low Internet connection was a preponderant factor in the 5 students in class 701 dropping out, as they were often unable to finalise one or two stages of the proposed activity; in addition, the early start of classes due to the absence of teachers on the day of the project contributed to discouraging the students, as under these conditions they would return home well before their normal departure time.

In Module 2, in both classes, the students took part in drawing up the mind map and in the fieldwork, even those who had dropped out of the previous module . One of the factors explaining the students' motivation and interest was probably the activities proposed, the drawing and the class held outside the school environment. It's important to emphasise that the questionnaire was administered to the residents and/or shopkeepers around the canal by the students taking part in the MMR.

In Module 3, both classes had greater participation in the MMR. In class 701 there was a 10 per cent increase, but this was due to new students joining the class; in class 702 there was a 4 per cent increase in involvement, due to a transfer from class 701.

A second aspect compared between the classes was the degree of motivation to use digital technologies to teach maps. In the first module, the multimedia record was close to zero, as only one student combined a photo with a map. In the third module there was greater student participation; 25% of class 701 produced videos and photos, while in class 702 there was greater student involvement, where 53% of students created media such as photos and videos that were later combined with maps.

Thirdly, we sought to analyse the students' level of knowledge in relation to basic cartography and geography content. In terms of the basics of cartography in teaching, both classes had difficulties, although it is not the norm to use all the basic elements of cartography (title, legend, scale, geographical coordinates) in a spatial representation.

We observed that the students had difficulties understanding projective and Euclidean spatial relationships, as well as notions of spatial orientation, which had not been properly worked on by their teachers in previous school years.

As a result, in the two Year 7 classes° , both in the mental maps and in the use of satellite images on Google Earth, as well as in the use of the GIS Web ArcGIS Online, the students had difficulties in giving a title to their map, identifying the graphic scale and, above all, creating a legend relating the objects on the map due to difficulties with vertical vision.

Fourthly, the MMR results in the two Year 7 classes° were compared on the basis of the students' performance in Geography during the 2013 school year.

Figure 63 shows the bimonthly averages of the two classes taking part in the research

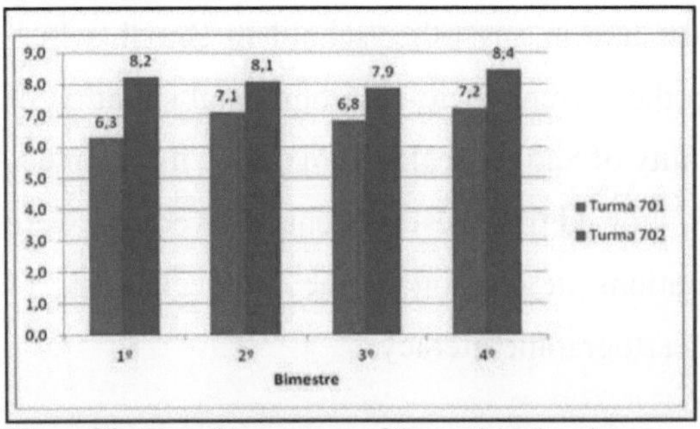

Figure 36 - Bimonthly averages of the participating classes

Source: Prepared by the author Sousa (2014)

With regard to class 701, the bimonthly average was 6.3 in the first bimonth (before the MMR was applied) and reached 7.2 in the fourth bimonth, an increase of 14 per cent, compared to class 702,

the average for the first two months was 8.2 and in the last two months (the fourth) there was a 2 per cent increase.

In both classes, among the multimedia resources combined with the maps produced in ArcGIS Online, texts were the media most used by the students as snippets of information about human interventions that caused the degradation of the Alcântara River channel, although to a lesser extent, the students also combined video or photo with their spatial representations.

Thus, the use of multimedia cartography for environmental education proved to be a potential technology for supporting geography classes. When analysed and compared the mind maps of both classes with the proposals to minimise and clean up the river channel, it was observed that this teaching methodology considered the students as subjects in the construction of knowledge through the use of photos, text and videos, and thus helped the students to correlate the influence of man on the environment, modifying and producing geographical space.

Certainly the use of geotechnologies connected to multimedia has helped them to improve their perception of their living space, but we detected the lack of another cartographic culture, just like in any other school in Brazil, which therefore made it difficult to grasp the current socio-environmental situation of the Alcântara River in the municipality of São Gonçalo, from one of its channels.

The application of the MMR showed that these students arrived and were approved for year 8° with limitations in relation to the basic elements of cartography and needed to resume cartographic literacy.

CHAPTER 5

CONCLUSION

Computer technologies and the Internet are increasingly used in the mapping process and this has led to greater circulation of maps and, above all, has expanded the possibility of producing these spatial representations in geography teaching.

The theoretical approaches of this research have made it possible to demonstrate the importance of multimedia cartography and geotechnologies as support materials for environmental education in cartography teaching. Teaching methodologies need to be developed so that multimedia resources and geotechnologies help to improve understanding and perception of geographical space.

In this sense, we understand the importance of developing teaching methodologies that take into account geographic and basic cartographic knowledge, according to the student's cognitive level, in order to include geotechnologies and multimedia resources.

The first stage of this research consisted of administering a questionnaire to geography teachers who teach seventh grade in the São Gonçalo/RJ municipal public school system, as well as to students in this year's classes at the Raul Veiga Municipal School.

We have seen that these teachers use paper maps from the textbook as cartographic materials, when they are not the only ones.

Teachers need to be aware of the possibilities and importance of using aerial photographs and satellite images as materials to support the teaching of cartography for the study of environmental dynamics, and not as "illustrations" taken from the textbook, as they mentioned. Other technologies applied to the mapping process, such as GIS, GPS, Google Earth, together with multimedia resources, can also be aids to cartographic language in Geography classes, as long as the methodologies are appropriate to the student's cognitive level in terms of basic cartographic knowledge and geographic reasoning.

Environmental education shouldn't just be worked on using ready-made maps found in geography textbooks; students need to be involved in spatial representation in order to understand geographical facts and phenomena based on their lived space.

Although all the teachers interviewed said that they deal with environmental issues in the municipality of São Gonçalo/RJ, contradictorily, the vast majority of them do not carry out any activities related to environmental issues; only two teachers out of a total of thirty carry out any cartographic activities related to the environmental problems of the surrounding area, in this case, the municipality or the Metropolitan Region of Rio de Janeiro.

It is necessary to overcome the dichotomy between geography teaching and environmental education so that social and environmental issues are understood in the interrelationships between society and nature in a single perspective for understanding the production of geographical space.

We found that there is a gap between the students, who are frequent users of computers and the Internet, and who have facilities for handling smartphones, and the pedagogical practices related to Environmental Education in Geography classes, since we did not detect the use of Multimedia Cartography, Geotechnologies, GPS, GIS, although we did minimally observe the use of Google Earth in cartographic language to work on environmental issues at different spatial and temporal scales.

It is important for teachers to be offered continuing education courses at extension or postgraduate level, covering the use of geotechnologies, multimedia resources and the Internet applied to cartographic language, aimed at developing environmental education in geography teaching. According to the answers to the questionnaire, most of these teachers do not use these new technologies, which can be obtained free of charge , such as GIS, satellite images and even aerial photographs, to work with up-to-date data on the environment, including data related to the space close to the student.

The evaluation of Mapping My River found that the students were initially reluctant to understand the current environmental state of the Alcântara River from one of its channels, whose waters they perceive as a ditch; the methodological proposal to use the computer, the Internet and activities outside the classroom in the schoolyard and fieldwork stimulated and motivated the students to use GPS, the Google Earth application and the ArcGIS Online web GIS when carrying out the activities proposed in the MMR modules.

The methodology of this investigation showed that the students were sent to the eighth grade with deficiencies in relation to the basics of cartography. We detected difficulties in interpreting the vertical view from satellite images, identifying geographical coordinates on GPS and applying the concept of scale. In addition, there were difficulties in coding and decoding the spatial information represented either on the maps produced with satellite images or even on the mental maps produced by the students themselves, dealing with the perception of the Alcântara River channel, which is part of their daily home-school journey.

It is worth emphasising that the choice of seventh grade for the MMR activities was based on the assumption that the students had already acquired and developed the basic knowledge to elaborate and read geographical space, taking into account the various forms of spatial representation that should have been learnt in sixth grade and in previous school years.

The use of mind maps in cartography teaching has proved to be a very valuable form of graphic representation, given the possibility for teachers to analyse, interpret, understand, learn about the readings and perception of geographical space, as well as to check their knowledge of the main elements of cartography, such as the legend, title and scale.

The results of the mind maps drawn up before the search for information, observations in the field and data collection on the socio-environmental perception of the Alcântara River showed a great concern with regard to the perception of geographical space from the student's lived place; the vast majority of mind maps were restricted to the side view of the Alcântara River channel. There was no concern to consider the spatial organisation around the river channel, or to correlate the main factors responsible for the pollution of its waters.

Despite the presence of disproportionate widths and lengths of the canal and between the objects represented next to the canal, difficulties with the notion of scale, together with the absence of natural and artificial objects, showed a lived space with little meaning, experience, affectivity and, above all, little geographical reasoning. For this reason, the use of mental maps is an important tool for working not only with the basic elements of cartography, but also with the social production of space and environmental issues. It should be emphasised that the purpose of the mind maps in this research was to find out how the students perceived the Alcântara River before the cartographic and geographical activities proposed in the MMR.

In general, the results of the application of the MMR activities showed that the ability to draw maps, read and interpret satellite images and understand the importance of basic cartographic elements throughout the development stages of the modules showed how cartographic literacy has been deficient.

Some of the contributions, limitations and possibilities of geotechnologies and multimedia resources in the understanding of geographical reasoning for Geography classes were analysed.

Among the contributions of the new digital technologies applied to cartography used in the development of MMR activities, the following stand out:

a) students' enthusiasm and interest in participating in cartography teaching. During the course of the MMR, the students were willing to carry out the stages of the activities in each module during free periods or even when the Geography teacher was absent; they often took a few minutes or even the whole break to finalise some stages or even the whole activity.

b) improved performance in Geography. Based on the arithmetic average of the four bimesters, it was found that in both classes the bimonthly grades increased, especially in class 701.

The limitations of the MMR were detected:

a) classes were brought forward due to a lack of teachers on the days the MMR was to be applied. This was undoubtedly one of the biggest limitations to the application of the material, as students were not released to carry out the MMR activities;

b) reduced number of computers in the computer lab. The classes were divided into two or three students per computer and separated into two large groups to take turns using the machines, as the eight computers could not accommodate the whole class;

c) The low Internet connection postponed the finalisation of the MMR. In addition, some stages of the activities had to be reworked, so the students had to do them outside of the MMR application time and even as homework;

d) the absence of an instructor in the Computer Lab during the course of the activities.

The application of the MMR in two seventh grade classes in the public school system showed possibilities for working with digital technologies applied to Cartography as support materials for the approach to Environmental Education in Geography classes, namely: a) carrying out tasks on environmental themes with up-to-date spatial data and information from the space close to the student;

b) The free use of GIS, satellite images, aerial photographs and Google Earth can make up for the school's lack of cartographic materials, such as thematic charts and maps, while having computers connected to the Internet;

c) consultation of the material outside the Geography class, in the school environment and in any environment that has a computer with internet access.

The results of the MMR tests in two seventh-grade primary school classes in the public school system showed that cartographic literacy has not been very productive or significant in terms of representing and understanding spatial reading, in relation to the use of legends, scale and vertical vision.

Despite the availability of other forms of graphic representation at no financial cost to the teacher and student, such as mind maps, GIS, satellite images and the use of the Google Earth application, teaching with ready-made maps from textbooks still predominates in São Gonçalo's municipal public schools.

It takes initiative and preparation on the part of the teacher to work with multimedia resources and geotechnologies as support materials for environmental studies in geography teaching, since the contribution of these technologies to the teaching and learning process requires methodologies that are appropriate to the student's cognitive level in terms of cartographic literacy and geographical knowledge.

Teachers can't get complacent and continue to teach cartography based only on printed and ready-made maps, restricting themselves to the use of textbooks amid the potential of multimedia cartography and geotechnologies. However, the Municipal Department of Education needs to reflect and rethink the conditions of the Computer Labs in the network's schools, so as to provide better infrastructure conditions so that teachers can exploit the potential of computer tools in the study of the environment through cartographic language.

The map, one of the forms of spatial representation, is not a representation of reality, but an interpretation of how the individual perceives the earth's surface, whether it is near or far. There's no point in developing teaching methodologies with multimedia resources and geotechnologies as support materials for Environmental Education if the teaching of cartography in Geography classes is done without the perspective of the student as the subject of the teaching and learning process. Both multimedia and geotechnologies do nothing to help locate, analyse and correlate the objects that make up geographical space if students have malformed spatial representations of their environment and of space beyond their direct experience.

In this way, we can see that carrying out fieldwork and using media such as photos, videos and texts to study the environment improves understanding of spatial dynamics. However, students need to have basic knowledge of cartography in order to create spatial representations and therefore improve their grasp of geographical space in a way that is more meaningful to the construction of abstractions based on their own reality, in other words, their lived space.

This methodology will be made available free of charge to the São Gonçalo/RJ Municipal Department of Education via a link to the MMR website as support material for the cartography teaching and learning process.

Future work related to the use of geotechnologies and multimedia resources will require further research in order to build methodologies and, consequently, improve the quality of teaching, both in Geography and in other subjects, such as Science, History, Maths, through an interdisciplinary approach to address socio-technical issues, bringing science and technology closer to the classroom: Science, History, Maths through an interdisciplinary bias to address socio-environmental issues bringing science and technology closer to the classroom.

To this end, in partnership with the city council, we intend to hold professional refresher courses for teachers and even for other education networks, both at municipal, state and federal level, showing the importance of developing and incorporating geotechnologies and multimedia into the study of geographical spatialisation as support material for their classroom practices in the construction of spatial knowledge.

Therefore, we hope that this methodology can serve as a model for future methodologies based on joint work involving Environmental Education and Cartography using Geotechnologies and multimedia resources, so that it becomes evident to develop and implement education mediated by geotechnology as an innovative and, above all, motivating and stimulating resource for both the teacher and the students.

REFERENCES

BACKER, Thomas. "WebGIS in Education". In: SOLARI, Osvaldo Muniz; DEMIRCI, Ali; SCHEE VAN DER, Joop. **Geospatial Technologies and Geography Education in a Changing World -** Geospatial Practices and Lessons Learned. Japan: Springer, 2015. p. 105-115.

BERNARDI, José Vicente Elias; LANDIM, Paulo. M. Barbosa. **Application of the Global Positioning System (GPS) in data collection.** Department of Applied Geology. Textbook. Rio Claro: Universidade Estadual Paulista Júlio de Mesquita Filho, 2002. 33 p. Available at <http://www.rc.unesp.br/igce/aplicada/ textodi.html>. Accessed on: 24 October 2013.

BOTELHO, Rosângela Guarrido Machado; SILVA, Antônio Soares da. Watershed and environmental quality. In: VITTE, Antonio Carlos; GUERRA, Antônio José Teirxeira. (Org.) **Reflexões sobre a Geografia Física no Brasil**. 5. ed. Rio de Janeiro: Bertrand Brasil, 2011. p. 153-192.

BRAZIL. **Constitution of the Federative Republic of Brazil**. Brasília, DF: Senate, 1988. 140p. Available at: <http://www.planalto.gov.br/ ccivil03/constituicao/constituicao.htm>. Accessed on: 26 August 2013

BRAZIL. Ministry of Education. **National Curriculum Parameters, Geography** (Third and Fourth Cycle of Primary Education). Brasília, DF, 1998. Available at: < http://portal.mec.gov.br/seb/arquivos/pdf/ geografia.pdf >. Accessed on: 8 January 2009.

BRAZIL. Ministry of Education. **National Curriculum Parameters, Transversal Themes** (Third and Fourth Cycle of Primary Education). Brasília, DF, 1998. Available at:<http://portal.mec.gov.br/seb/arquivos/ pdf/transversais.pdf>. Accessed on: 12 October 2010.

BRAZIL. **National Policy on Environmental Education**. Official Gazette Federative Republic of Brazil, Brasília, DF, 28 Apr. 1999. Available at: <http://www.planalto.gov.br/ccivil_03/leis/l9795.htm>. Accessed on: 10 Feb. 2013Callai (2013)

CAPEL, Horacio. Perception of the environment and geographical behaviour. **Revista de Geografia**, v.7, n.1-2, p. 58-150, 1973. Available at:

<http://www.raco.cat/index.php/<RevistaGeografia/ article/viewFile/ 45873/56665..10/10/2013>. Accessed: 10 Oct. 2013.

CARLOS, Ana Fani Alessandri. Defining Place? In: _____ . **The place of/in world**. São Paulo: Labur Electronic Edition, 2007. p. 17-20.

CARTWRIGHT, William. Development of Multimedia. In: ____; CARTWRIGHT, William; PETERSON, Michael P.; GARTNER, Georg. **Multimedia Cartography**. Berlin: Springer Verlag, 1999. p.11-30.

CRAMPTON, Jeremy W. Maps 2.0: map mashups and new spatial media. In: ____. **Mapping:** A Critical Introduction to Cartography and GIS. United Kingdom: John Wiley & Sons, 2010. p. 13-24.

CASTELLAR, Sônia Maria Vazella. Genetic psychology and learning in geography teaching. **Geographical education:** theories and teaching practices. São Paulo: Contexto, 2005. p. 38-50.

CASTELLAR, Sônia Maria Vanzella. Cartography and the construction of knowledge in the school context. In: ALMEIDA, Rosângela Doin de (Org.) **Novos rumos da Cartografia escolar** - currículo, linguagem e tecnologia. São Paulo: Contexto, 2011. p. 121-135.

CASTRO, Iná Elias. The problem of scale. In: ______. GOMES, Paulo César; CORRÊA, Roberto Lobato. **Geography:** Concepts and Themes. 2. ed. Rio de Janeiro: Bertrand Brasil, 2000. p. 117-140.

CAVALCANTI, Lana de Souza. Geographical science and geography teaching. In:. **Geography, School and the Construction of Knowledge**. 9. ed. Campinas: Papirus, 2006. p. 17-28.

DEBESSE-ARVISET, M-L. **The School and Environmental Aggression**. São Paulo: Difel, 1974.

DI MAIO, Angelica Carvalho. **Digital Geotechnologies in Secondary Schools:** a practical evaluation of their potential. 2004. 189 f. Thesis (PhD in Geography) - Institute of Geosciences and Exact Sciences, Universidade Estadual Paulista Júlio de Mesquita Filho, Rio Claro, 2004.

DI MAIO, Angelica Carvalho; SETZER, Alberto Waingort. Education, Geography and the challenge of new technologies. **Revista Portuguesa de Educação**, Braga, v. 24. n.2, p. 211-241, 2011.

FLORENZANO, Teresa Gallotti. Remote Sensing as a teaching resource. In: ______________. **Initiation in Remote Sensing**. Images of

Satellite for Environmental Studies. 2. ed. São Paulo: Oficina de textos, 2007. p. 95-101.

FONSECA, Fernanda Padovesi; OLIVA, Jaime. **How I teach Cartography**. São Paulo: Melhoramentos, 2013.

FREUNDSCHUH, Scott M.; HELLEVIKS, Wesley. Multimedia Technology in Cartography and Geographic Education. In: CARTWRIGHT, William; PETERSON, Michael P.; GARTNER, Georg. **Multimedia Cartography**. Berlin: Springer-Verlag, 1999. p. 271-280.

GARDNER, H. **Multiple Intelligences**: Theory in Practice. Artes Médicas: Porto Alegre, 1995.

GASPAR, José; MARIAN, Ana. The perception of space. **Finisterra - Revista Portuguesa de Geografia**, Lisbon, v.20, n. 20, p. 319-322, 1975. Available at:< http://www.ceg.ul.pt/finisterra/numeros/1975-20/20_10.pdf>. Accessed on: 19 October 2013.

GOMES, Nulo Filipe Lopes. **The didactic potential of Geographic Information Systems in Geography Teaching and its application to the 3rd cycle of Basic Education**. 2006. 159 f. Dissertation (Master's Degree in Geographic Information Science and Systems) - Instituto Superior de Estatística e Gestão de Informação, Universidade de Nova Lisboa, Lisbon, 2006. Available at: <http://labnt.isegi.unl.pt/links_CV/ potencial_did%C3%Actico_dos_SIG_no_ensino_da_geografia.pdf.> Accessed on: 22 Apr. 2009.

GOMES, Paulo César. The counter-currents. In:______. **Geography and Modernity**. Rio de Janeiro: Bertrand Brasil, 1996. p.93-124.

GONÇALVES, Amanda Regina; ANDRÉ, Iara Regina Nocentini; AZEVEDO, Thiago Salomão; GAMA, Valquíria Z. Analysing the use of **"Google Earth"** images **and maps in Geography teaching. Ar@acne - Revista Electrónica de Recursos em Internet sobre Geografía y Ciencias Sociales**, Barcelona, n. 97, Jun. 2007. Available at:< http://www.ub.edu/geocrit/aracne/aracne-097.htm>. Accessed on: 10 Jan. 2012.

GOULD, Peter; WHITE, Rodney. The images of places. In: ______ . **Mental maps**. 2. ed. London: Taylor & Francis, 2002, p. 12-16. E-book.

HONG, Jung Eun. **Web-based GIS for middle school teachers:** using online mapping applications to promote teacher adoption. 2012. Thesis

(Doctor of Philosophy) - Faculty of the Graduate School, University of Colorado at Boulder, Colorado.

KENSKI, Vania Moreira. Reflections and questions about the digital society and the formation of a new professional/teacher. **Revista Latinoamericana de Tecnologia Educativa-RELATEC**, v. 3, n. 2. p. 99-107, 2004. Available at: < http://campusvirtual.unex.es/ /revistas/index. php/relatec/article/view/171>. Accessed on: 29 September 2009.

KERSKI, Joseph J. **Exploring the World in 3D with Google Earth:** Could it be that the world is finally discovering how useful geography really is? 2006. Available at:<http://education.usgs.gov/comum

/.../google_terra_in_education.pdf>. Accessed on: 25 May 2009.

KERSKI, Joseph. "Understanding our changing world through web-mapping based investigations.". Journal of Research and Didactis in Geography, n. 4, p. 1- 13, Oct. 2013. Available at:<http://www.j-reading.org/ index.php/geography/article/ view/39.> Accessed on: 06 July 2016.

KOZEL, Salete. Mind maps - A form of Language: Methodological Perspectives. In: ______SILVA, Josué da Costa; GIL FILHO, Sylvio Fausto. **From Perception and Cognition to Representation:** Theoretical Reconstructions of Cultural and Humanist Geography. Curitiba: Terceira Margem, 2007. p.114-138.

KOZEL, Salete. Representation of space from the perspective of concepts: lived world and dialogism. In: ENCONTRO NACIONAL DE GEÓGRAFOS, 16., 2010, Porto Alegre. Crisis, praxis and autonomy: spaces of resistance and hope - Space for the Socialisation of Collectives, 2010. Porto Alegre, 2010. Available at:< http://www.agb. org.br/evento/download.php?id Trabalho=4528> Accessed on: 20 Oct. 2013.

KRAAK Menno-Jan. Trends in Cartography. In ______.; BROWN, Allan. **WEB Cartography:** developments and prospects. New York: Taylor & Francis, 2001, p. 9-19.

LÉVY, Jacques. A cartographic turn? In: ACSELRAD, Henri (Org.) **Social cartographies and territory**. Rio de Janeiro: UFRJ/IPPUR, 2008. p. 153-167.

LÉVY, Pierre. **Cyberculture**. São Paulo: Editora 34, 1999.

LUSSAULT, Michel. Image. In: LÉVY, Jacques; LUSSAULT, Michel. **Dictionnaire de la Géographie et de 1'espace des sociétes.** Translated by

work by Fernanda Padovesi Fonseca and Jaime Tadeu Oliva. Paris: Belin, 2003. p. 485-489.

MARTINELLI, Marcello. Environmental cartography: a different kind of cartography? **Revista do Departamento de Geografia**, São Paulo, v.26, n.7, p. 61-80. 1994. Available at:<http://citrus.uspnet.usp.br/<rdg/ojs/ index.php/rdg/article/view/237/0>. Accessed on: 10 June 2013.

MARTINS, Luciana Junqueira. SEABRA, Vinícius da Silva; CARVALHO, Vânia Salomon Guaycuru de. The use of Google Earth as a tool in basic geography teaching. In: Brazilian Symposium on Remote Sensing, 16. **Anais** ..., 2013, Foz do Iguaçu. Available at:< http://www.dsr.inpe.br/sbsr2013/files/p0251.pdf >. Accessed on: 20 January

2013.

MENEZES, Paulo Márcio Leal de. **New cartographic technologies in support of geotechnologies and geographical analysis**. Available at:<http://www.geocart.igeo.ufrj.br/pdf/trabalhos/2003/ Cartografia_e_Novos_Paradigmas_2003.>. Accessed on: 16 July 2009.

MOREIRA, Sueli Aparecida Gomes. **Multimedia cartography:** interactivity in cartographic projects. 2010. 123 f. Thesis (PhD in Geography) - Institute of Geosciences and Exact Sciences, Universidade Estadual Paulista Júlio de Mesquita Filho, Rio Claro, 2010. Available at:< http://acervodigital.unesp.br/handle/123456789/61149>. Accessed on: 3 Feb. 2012.

NASCIMENTO, Rosemy da Silva. **Tools for formal environmental education with a focus on water resources**. 2003. 240f. Thesis (Doctorate in Environmental Management) - Technology Centre, Federal University of Santa Catarina, Florianópolis, 2003.

NOGUEIRA, Amélia Regina Batista. Mind map: didactic resource for the study of place. In: PONTUSCHKA, Nídia Nacib.; OLIVEIRA, Ariovaldo Umbelino de (Org.). **Geografia em perspectiva:** ensino e pesquisa. 3. ed. São Paulo: Contexto, 2010. p. 125-133.

NOGUEIRA, Ruth Emília; DOMINGUES, Júlio; CHAVES, ANA Paula Nunes. A website for schoolchildren: interactive content for learning cartography. In: Colóquio de Cartografia para crianças e escolares e II Fórum Latino-americano de Cartografia para escolares, 6., **Anais ...** 2009, Juiz de Fora: UFJF, 2009. Available at: <http://www.labtate.ufsc.br/images/A3.pdf>. Accessed on: 15 March 2012.

NOVAES, André Reyes. Favelas and the divided city: mapping silences and calculations in Rio de Janeiro's journalistic cartography. **Social & Cultural Geography**, v.15, London: Taylor & Francis, 2014. p. 201-225.

NUNEZ, José Jesus Reyes. **A GIS site for Hungarian pupils**. Available at: < http://www.datamap-bg.com/conference_cd/pdf/; 13_208_Jesus _Hu.pdf>. Accessed on: 25 Feb. 2014.

OLIVEIRA, Lívia. **The methodological and cognitive study of the map**. 1977. 203 f. Tese (Livre-docência)- Instituto de Geociências e Ciências Exatas, Universidade Estadual Paulista Júlio de Mesquita Filho, Rio Claro, 1977.

PAZINI, Dulce Léia Garcia. Using Geoprocessing Technologies in Geography Teaching: a methodological proposal for primary education (3rd and 4th cycles). Available at: <http:// www.inpe.br/unidades/ cep/atividadescep/

jornada/programa/t-11_trab_17.pdf>. Accessed on: 02 May 2014.

PERRENOUND, Phillipe. Using new technologies. In: ______ . **Ten new competences for teaching**. São Paulo: Artmed, 2000. p.125-141.

PETERSON, Michael P. Elements of Multimedia Cartography. In: CARTWRIGHT, William..; ___ .; GARTNER, Georg. **Multimedia Cartography**. Berlin: Springer Verlag, 1999. p. 31-40.

PETERSON, Michael P. That Interactive Thing You Do. **Cartographic Perspectives**, n. 29. p. 3-4.1998. Available at:<http://maps.unomaha. edu/mp/Articles/ThatInteractive.html>. Accessed on: 22 July 2011.

PETERSON, Michael P. The Internet and Multimedia Cartography. In: CARTWRIGHT, William; PETERSON, Michael P.; GARTNER, Georg. **Multimedia Cartography**. 2. ed. New York: Berlin: Springer-Verlag. 2007. p. 35-48.

PONTUSCHKA, Nídia Nacib.; PAGANELLI, Iyda Tomoko.; CACETE, Núria Hanglei. Study of the environment: significant moments of apprehension of reality.
In: _____ . **To teach and learn Geography**. São Paulo: Cortez, 2007. p. 171- 212.

PORCHER, Louis; FERRANT, Pierre; BLOT, Bernard. **Pedagogy of the environment**. Lisbon: Socicultur, 1975.

RAMOS, Cristhiane da Silva. Multimedia. In:. **Cartographic visualisation and multimedia cartography - Concepts** and Technologies. São Paulo: UNESP, 2005. p. 49-70.

REIGOTA, Marcos. **What is environmental education?** 2. ed. Rio de Janeiro: Brasiliense, 2009.

RICHTER, Dennis. Teaching Geography and Mind Maps. In: ______ . **O mental map in Geography** teaching - Conceptions and proposals for teaching work. São Paulo: Cultura acadêmica, 2011. p. 97-132. Available at:<http://www.culturaacademica.com.br/_ img/arquivos/ O_ mapa_mental_no_ensino_de_geografia.pdf>. Accessed on: 28 September 2014.

SACHS, Ignacy. **Paths to sustainable development**. 4. ed. Rio de Janeiro: Garamond, 2002.

SANTAELLA, Lúcia. From media culture to cyberculture: the advent of the post-human. **Famecos Magazine**. Porto Alegre, December 2003. Available at:

<http://revistaseletronicas.pucrs.br/ojs/index.php/revistafamecos/ article/viewFile/3229/2493>. Accessed on: 20 Feb. 2013.

SANTOS, A. A. S. **Geographic Information Systems in geography teaching:** application to a 3rd cycle primary school class. 2010. 96f. Master's dissertation. Instituto Superior de Estatística e Gestão de Informação, Universidade Nova de Lisboa, Lisbon, 2010. Available at: <http://run.unl.pt/bitstream/10362/5365/1/TEGI0270.pdf>. Accessed on: 20 Feb. 2014.

SANTOS, Milton. Structure, process, function and form as categories of the geographical method. In: ____. **Space & Method**. São Paulo: Nobel, 1985. p. 49-59.

SAO GONÇALO. **Participatory Municipal Masterplan**. São Gonçalo, RJ. 2006. 44 p. Available at:<http://www.pmsg.rj.gov.br/ urbanismo/plano_diretor/leitura_tecnica/>. Accessed on: 7 May 2012.

SAO GONÇALO. **Law n° 016, of 11 July 2011**. Provides for the Environmental Policy of the Municipality of São Gonçalo. São Gonçalo City Council. São Gonçalo, RJ. 11 July 2001.

SAUVÉ, L. A cartography of currents in environmental education. In: SATO, M.; CARVALHO, I. C.M. Educação Ambiental. Porto Alegre: Artmed, 2005. p. 17-45. Available at:< http://www.institutoaf.org.br/ Sato_artmed%5B1%5D.pdf>. Accessed on: 15 September 2013.

SEEMANN, Jorn. Maps and environmental perception: from the mental to the material and vice versa. **OLAM. magazine**, Rio Claro, v. 3, p. 200-223, Sep. 2003.
Available at:<http://www.academia.edu/187819/ Maps_and_

percepcao_ambiental_do_mental_ao_material_e_vice-versa>. Accessed on: 03 October 2013.

SANTOS, Vânia Maria Nunes. The school use of Remote Sensing as a pedagogical teaching resource in the study of the Environment. In: **COURSE ON THE USE OF REMOTE SENSING IN THE STUDY OF THE ENVIRONMENT**. São José dos Campos, 2002, p. 1-15. Available at:<http:// mtc-m12.sid.inpe.br/col/Cap 12-_VMNSantos.pdf>. Accessed on: 17 Feb. 2009.

SILVA, M. **Sala de aula interativa**. 3. ed. Rio de Janeiro: Quartet, 2001.

SIMIELLI, Maria Elena Ramos. Cartography in primary and secondary education. In: CARLOS, Ana Fani Alessandri (Org). **Geography in the classroom**. 5. ed. São Paulo: Contexto, 2003. p. 92-108.

SOUZA, José Cavalcante (Org.) **Os pré-socráticos:** fragmentos, doxografia e comentários. São Paulo: Nova Cultural, 1996.

STASZAK, J-F. Entry "Mental map" - Cartographic expression of a subjective representation of space. Translated by Fernanda Padovesi Fonseca and Jaime Tadeu Oliva. In: LÉVY, Jacques; LUSSAULT, Michel. **(Dictionnaire de la Géographie et de Kespace des societies**. Paris: Belin, 2003.

TAYLOR, D R Fraser; LAURIAULT, Tracey P. Future Directions for Multimedia Cartography. In: CARTWRIGHT, W. PETERSON, M, P; GARTNER, G. **Multimedia Cartography**. Berlin: Springer Verlag, 2007. p. 505-522. Available at: <http://www.interpares.org/display_file.cfm?doc=ip2_dissemination_bc_taylor~lauriault_2006.pdf>. Accessed on: 20 Feb. 2013

TAYLOR, D R Fraser. The theory and practice of Cybercartography: An Introduction. In: _____ . **Cybercartography:** Theory and Practice. Amsterdam: Elsevier, 2005. p. 22-34.

TEIXEIRA, Kozel Salete; NOGUEIRA, Amélia Regina Batista. The Geography of representations and its pedagogical application: contributions from a lived experience. In: **Revista do Departamento de Geografia**, São Paulo, SP, n.13, p. 239-257, 1999. Available at:< http://citrus.uspnet.usp.br/rdg/ojs/index.php/rdg/article/view/158/138>. Accessed on: 20 Dec. 2013.

TRISTÃO, Martha. **A Educação Ambiental na formação de professores -** rede de saberes. 2. ed. São Paulo: Annablume, 2008

TUAN, Yi-FU. Topophilia and the environment. In: . **Topophilia -** A study of the
perception, attitudes and values of the environment. São Paulo: Difel, 1980.p. 106-128.

VILHENA, Drielly Costa; TAVARES, Stênio Soares; NETA, Luiza Câmara Beserra. Remote sensing as a didactic resource for teaching Geography. In: **Revista Geonorte**, Manaus, AM, n.4, p. 1624-1635, 2012. Available at: <http://www.revistageonorte. ufam. edu.br/ attachments/009 (O%20 SENSORIAMENTO%20REMOTO%20COMO% 20RECURSO%20DID%C3%81TICO%20NO%20ENSINO%20DA%20GEOGRAFIA).pdf>. Accessed on: 15 April 2014.

WIEGAND, Patrick. Practical map activities: age 11 to 14. In: _____ . **Learning and Teaching with maps**. New York: Routledge, 2006. p. 106115.

WIEGAND, Patrick. **The best of both worlds?** Complementary in educational cartography. Available at: <http://lazarus.elte.hu/cet/ madrid-2005/wiegand.pdf>. Accessed on: 10 Nov. 2010.

WOLFGRAM D. E. **Creating in Multimedia**. Rio de Janeiro: Campus, 1994.

More Books!
yes

MIX
Papier aus verantwortungsvollen Quellen
Paper from responsible sources
FSC® C105338

Printed by Books on Demand GmbH, Norderstedt / Germany